Ethics and Practice in Science Communication

Ethics and Practice in Science Communication

EDITED BY SUSANNA PRIEST, JEAN GOODWIN,
AND MICHAEL F. DAHLSTROM

The University of Chicago Press
Chicago and London

The University of Chicago Press, Chicago 60637
The University of Chicago Press, Ltd., London
© 2018 by The University of Chicago
Published 2018
Printed in the United States of America

27 26 25 24 23 22 21 20 19 18   1 2 3 4 5

ISBN-13: 978-0-226-54060-3 (cloth)
ISBN-13: 978-0-226-49781-5 (paper)
ISBN-13: 978-0-226-49795-2 (e-book)
DOI: https://doi.org/10.7208/chicago/9780226497952.001.0001

Library of Congress Cataloging-in-Publication Data

Names: Priest, Susanna Hornig, editor. | Goodwin, Jean, editor. | Dahlstrom,
    Michael F., editor.
Title: Ethics and practice in science communication / edited by Susanna Priest,
    Jean Goodwin, and Michael F. Dahlstrom.
Description: Chicago ; London : The University of Chicago Press, 2018. |
    Includes bibliographical references and index.
Identifiers: LCCN 2017031741 | ISBN 9780226540603 (cloth : alk. paper) |
    ISBN 9780226497815 (pbk. : alk. paper) | ISBN 9780226497952 (e-book)
Subjects: LCSH: Communication in science—Moral and ethical aspects.
Classification: LCC Q223 .E74 2018 | DDC 174/.95—dc23
LC record available at https://lccn.loc.gov/2017031741

Contents

Foreword

With so much attention given to how to improve science communication, it is remarkable that consideration of the ethics of science communication has been so lacking. This book is an important step to correct that lack.

Scientists and commentators long have been concerned with ethics in the practice and application of scientific research. For example, practices such as the abuse of human research subjects or the use of unsafe lab equipment, putting research assistants at risk, have led to new standards imposed on scientists. Occasional fraud in science has led to concerns about selfish motivations and the weaknesses in peer review. Other examples abound of ethical considerations in the practice of science.

So, too, there has been much attention to the ethics of technology. Perverse misuse of powerful technologies derived from research has harmed or oppressed people and has led to much ethical commentary. Volumes have been written about the effects of technology on individuals and on society.

Although discussion will continue about what is appropriate ethical behavior in conducting scientific research and in applying technologies, there is wide agreement that they are drenched in ethical considerations. But what about communication? Does writing or talking with others about scientific ideas carry ethical implications, as well?

Yes, and the recognition that ethics is inherent in communication about scientific ideas is much overdue. Scientists tend to communicate in ways and for reasons that they presume to be ethical. Up to now, most discussion, at least among scientists, about the communication of science has focused on the techniques to get the message across clearly. If the audience is other scientists, the emphasis has been on getting the message across fully so that colleagues can understand the collection and analysis of evidence well enough to determine whether they can come to the same conclusions as the authors. For communication with the public, the emphasis has been on simplification, finding the minimal set of facts that will allow an untrained person to appreciate the scientific idea at hand. Communicators who can enhance clarity with the addition of drama, narrative, or even poetry sometimes gain praise from other scientists, although more often they receive opprobrium for "not sticking to the facts." In any case, little attention is given to who the audience might be, what the communicator owes the audience, and what else should be communicated beyond the "simple facts."

Of course, clarity is essential. But clarity about what? Does the writer or speaker do the audience and even society at large a disservice if, say, he or she fails to explain the long-term possible effects of the findings on society, or fails to explain the possible risks to specific populations, or fails to specify the next research steps necessary for results to become more than a one-time curiosity? We have few guidelines or frameworks in which to discuss these ethical issues. Communication training, much less training regarding the ethics of communication, is rarely part of formal science training, while training on the ethics of the scientific process is often a requirement of funding awards and a component of graduate training. This collection begins to fill that gap by examining a variety of situations that demonstrate that a scientist has important ethical obligations beyond operating a safe, honest, and productive research group and preventing misuse of technologies to harm people. Some of these obligations involve communication.

Science is special. It has been demonstrated over the centuries to be the means for acquiring public knowledge that is the most reliable, most resilient, and most applicable to public problems. Indeed, the scientific way of thinking could be called the world's greatest civilizing influence over the past half millennium. Because the process is so powerful, it naturally comes with ethical obligations to do it right. There are obligations that the practice, the communication, and the application of science be conducted with integrity and human compassion. Inadequate or improper communication of science has consequences.

Communication is essential for science to work at all. Science is a rubric for asking questions so that they can be answered empirically and verifiably. Essayist and physician Lewis Thomas memorably called science "the shrewdest maneuver" for understanding how the world works. A scientist owes other scientists clear communication so that the work can be checked, the understanding can be refined, and progress can be made. Without communication in a manner useful to other scientists, the work is not science but simply personal indulgence. That ethical obligation to contribute to human progress, although sometimes ignored, is well established. Misleading other scientists through omission of essential information or exaggerating the quality and importance of evidence or conclusions is not simply bad form but a violation of scientific ethics. The discussions in this book go further. Deeper reflection leads to the realization that because science takes place in a societal context, failures of commission or omission, whether involving popular or scholarly communication, are ethically compromising. Part of the reason is that scientists have a heavy responsibility to support nonscientists in our self-governing democratic society. When we consider this heavy responsibility, the absence of more books like this one is noteworthy.

Science is fundamentally a human enterprise. Public communication exposes the tension between the ideal of an objective scientist, who is seen to have great credibility and authority, and the more human reality of a scientist who is embedded in human social institutions—and in society itself.

Thus, it is surprising that for many, many years most studies of science communication have been directed toward techniques for improving the effectiveness of the communication without direct consideration of the ethics. Yet there are many subtleties involved in communicating science that not only hamper or enhance the effectiveness of communication, but also determine whether a speaker or writer can meet their responsibilities to an audience.

For more than a century and a half, the American Association for the Advancement of Science (AAAS), the world's largest general science membership organization, has been concerned with effective science communication. For more than a decade, the AAAS Dialogue on Science, Ethics, and Religion has explored ways that scientists, faith leaders, and religion writers can strengthen each other's understanding. AAAS EurekAlert! provides thousands of prepublication announcements of science advances, giving journalists material for their writing. Each year the AAAS Science and Technology Policy Fellows program trains 280 PhD scientists to bring a scientific perspective for a year to legislative and executive offices. A variety of programs for undergraduate and graduate students bring insight to the students, both about communication and about understanding the research enterprise itself. In recent years the AAAS Office of Public Engagement has taken part in the growing field of "the science of science communication," looking at evidence-based strategies for the practice of public communication. Ethical considerations have begun to slip into these and other communication activities at AAAS. However, more explicit and specific attention to communication ethics, such as this collection provides, will help.

Consideration of the ethical implications of communicating science also leads to more careful consideration of precisely who is the audience and what is owed to them. It is interesting to think that perhaps one effect of a book like this, intended to advance ethics, may also be to lead scientists and others to pay more attention to the audience and hence to improve the understandability of their communication. Of course, getting scientists and other science communicators to pay more attention to the many ethical considerations involved in science communication is the more central purpose of this book.

Traditionally, scientists have assumed their responsibility to be to present the facts accurately. Hardly ever does the scientist consider what is meant by "accurately" or even by "facts." Does the reader or listener have the same definitions as the speaker? Some years ago when one of us (R. H.) was in an orientation session for AAAS science policy fellows who were about to begin a year as staff members on Capitol Hill, one speaker left the scientists unsettled by telling us, "You must understand that here in Washington, facts are negotiable." Later we came to understand that part of his message was that there are two sides to the communication of scientific knowledge. Any scientists who think they have fulfilled their obligation by feeding simplified facts to a policy maker, in one direction only, may not only have failed to communicate effectively, but may also have done real damage to the policy-making process.

As the contributors to this book present clearly and provocatively, anyone who attempts to communicate science, whether a scientist or a science communicator, enters into an ethical contract with the audience. It is a contract that, if executed properly, requires establishing a trust between the audience and the communicator, determining the degree of certainty and authority the writer or speaker will bring to the subject, and insisting on transparency about the implications of the science and of any action being advocated. As the editors write in the introduction to part III, "Being an accomplished and ethical science communicator requires more than knowledge of the science itself or of ways to render it more readily understandable. It requires considerable social, political, and cultural awareness."

Science has such great power to produce clear thinking about the world and to alleviate human ills that any communication that does not maximize the benefits is ethically compromised. Citizens deserve the opportunity to understand science and to make evidence-based decisions. The need for nonscientists to appreciate the essence of science and use evidence-based thinking for themselves is greater than the need to understand the details, terminology, or methodology of any particular scientific experiment or concept. Any public communication about any scientific subject is blameworthy if it does not address at least partially this central problem facing society today: how citizens can make decisions about public issues based on evidence— that is, how all citizens can think like scientists, even if not themselves scientists by profession.

Rush D. Holt, chief executive officer, AAAS, and executive publisher, Science
Jeanne Braha, former project director, public engagement, AAAS

Introduction to This Book

SUSANNA PRIEST, JEAN GOODWIN,
AND MICHAEL F. DAHLSTROM

Why do we need an entire book on the ethics of science communication? It might seem as though a fairly obvious set of principles is involved, shared by both journalists and scientists. Be accurate and get the science right. Strive to communicate clearly so nonscientists can understand. Do not falsify or distort, do not plagiarize. Do identify sources, and do acknowledge competing views.

All of these things are involved, of course, and all are important. However, these common and reasonably well-recognized principles do not exhaust the wide universe of ethical challenges that science communicators face in everyday practice. In fact, ethical questions permeate many more aspects of the work of science communicators than might at first appear to be the case. Our central goal in creating this book has been to encourage thinking beyond the basics. We seek to stimulate "reflexive" thinking—thinking about the broader consequences of our own actions, including the actions involved in communication—so that we can begin to better define and articulate the ethical dimensions of those actions.

Reflexivity is especially important in the context of contentious issues in which the science-society relationship has become fraught with tension. When this happens, very often issues of underlying worldviews (values and beliefs) are in play, not just issues of knowledge. We also wanted to bring different kinds of scholarship to bear on this thinking, primarily through including the voices of scholars from diverse communication fields, but also voices from different branches of philosophy. The views of scientists themselves are also considered. (In particular, chapter 9 by Davies analyzes in detail a group of scientists' perspectives on science communication ethics, with interesting conclusions.) We also included consideration of worldviews that deviate from those of so-called Western science (see chapter 14 by Coleman).

Beyond the Basics

Science communicators come from many backgrounds and work toward different—sometimes competing—objectives. Journalists; outreach, extension, and museum specialists; and public information officers (among others) are charged with disseminating research results, often distilled within the walls of the university, to people outside those walls who need or want to make use of them. This dissemination takes many forms, and often goes beyond simply transmitting facts. Sometimes the communicator tries to excite the audience, perhaps by adding in an ingredient of wonderment or mystery; sometimes the communicator may even aim to entertain. Public relations professionals working with science use many of the same skills and strategies but have a different mission: They are explicitly tasked with promoting something or someone, advancing the interests of advocacy organizations or corporations, or managing the reputations of research institutions. A whole host of government agencies also have missions that include creating, interpreting, and communicating valid and reliable scientific information. All of these people can be described as science communicators—it's a vast territory.

And then, of course, there are scientists themselves, who may not think of themselves primarily as science communicators but who are called on quite regularly to communicate their work to other scientists, to government officials, to journalists and other communication professionals, or to any of a wide range of non-specialist audiences—from family and friends to strangers on an elevator, from participants in science-oriented public events to other, more general, kinds of community groups. For many scientists, science communication may seem more like a skill set than a field of study. But there is more to it than meets the eye, even so.

These diverse science communicators are going to run into diverse practical problems. Their audiences are going to vary. Some will be very interested in new scientific results, such as the people who turn up at science museums or who read the science pages of newspapers. Others will be less interested. A few will be misinformed about science, and an even smaller group actively hostile to scientific points of view. The science being communicated will also be diverse, ranging from established theories backed by a scientific consensus to new results the significance of which is subject to a high degree of disagreement. And there will always be uncertainties.

Given the complexities of communicator, audience, and the science to be communicated, it is no wonder that communicating science *effectively* is a challenging task. In this volume, we are not going to make the process any simpler; indeed, we want to deepen our understanding of its complexity by opening questions of the additional *normative* dimensions of science communication.

Human beings are remarkable in terms of the unique complexity of the societies they organize, and every human society has a complex set of behavioral expectations for its members. Only a few of these are actual rules that can be codified into formal law or articulated as specific requirements or prohibitions. The rest may still fall under what social scientists generally think of as norms or normative expectations for behavior, a broader category.[1] We are very conscious of even minor deviations from these expectations—when someone behaves in an "odd" way, for example, our internal radar is quickly alerted to the transgression. Yet norms are often implicit rather than explicit—perhaps we cannot even put our finger on what is "odd" when one of these is broken. Many social scientists argue that norms are never really fixed but constantly emerge and evolve in response to particular contexts. Learning the norms of a new culture or subculture—say, when a graduate student in science first encounters normative questions surrounding peer review—can involve a steep curve.

Ethics and morals can both be seen as subsets of our normative expectations. For some, these subsets overlap quite a bit. Ethical rules may be seen as reflecting norms for which it is generally and explicitly recognized that deviating behavior is *morally* wrong—it is "bad" behavior. Others may make a much stronger (if varying and often subtle) distinction between ethics and morals, rather than conflating the two as we tend to do in casual conversation. Our goal here, however, is not to sort out the longstanding debate over this distinction, but only to highlight that the terms are not always used interchangeably—or even consistently. What constitutes moral or immoral—and ethical or unethical—behavior can be defined differently by different people (and in different cultures), and the terms themselves are used in different ways even among specialists. Some ethical ideas (but not all) are not

---

1. While norms should not be understood as a kind of statistical average of what we observe other people doing—that is, it is not generally okay to excuse a "bad" behavior by arguing that "everyone is doing it"—norms do change over time and according to circumstances, and they may also be affected by our observations of others. (Think, for example, of dress codes, whether written into formal rules or existing simply as everyday expectations that are "understood." These norms are highly context-specific and constantly evolving.) When we go into any new cultural situation, we "pick up on" how others are behaving and use that information to adjust our own behavior, a process that social scientists call "socialization."

written into laws; others are written into codes for professional conduct—or accepted as professional standards even if neither law nor code. It seems this difficult-to-define concept of ethical behavior is an inherent part of what it means to be human. Indeed, the worst sorts of unethical (and immoral) behavior are described as "inhumane." (For more about how the way the word "ethics" is used, see chapter 2 by Thompson.)

Today, our challenges in science communication arise from an ever-expanding range of new communicators, new audiences, and new science. Of course, new ethical issues also arise from the use of new communication channels such as blogs, tweets, and online discussions to talk about science, further complicating things. None of this seems permanently fixed; we have grown accustomed to an environment in which both science and technology are constantly emerging. So are the forms that science communication should take. And whenever that "should" word comes into play, we can recognize that an ethical issue is most likely at stake.

Beyond the Deficit Model

In addition to moving beyond the ethical basics, in this volume we have also sought to move beyond some common but often inadequate assumptions about science communication. This means primarily moving beyond what is often called "deficit model" thinking. Several of the papers in this volume take their departure from this so-called model of science communication, so a brief explanation of the idea should be useful. Rarely put forward as an actual, explicit theory, the term "deficit model" is intended to capture the generally *implicit* assumption (implicit in the same way many social norms are implicit—that is, taken for granted) that public resistance to science is largely attributable to a lack of scientific knowledge or science literacy. This is a "common sense" assumption for many scientists and science communicators, but empirically it is not always accurate.

In deficit model thinking, if some scientific result or technological innovation is not being taken up by the public, or if science in general does not seem to be receiving enough public support, increasing scientific knowledge through improved science communication is seen as the answer. The root cause of the perceived problem is identified as the public's lack of understanding, so the solution is to provide them with better information—filling the deficit in their knowledge. Science communicators are (in this view) people with access to the kind of knowledge that needs to be transmitted in order to solve the public perception "problem." Unfortunately, the result is often an attempt to transmit the requisite knowledge in an authoritarian, top-down fashion—one that can too easily ignore real public concerns rather than addressing them. The audience is likely to be seen as a passive, undifferentiated "mass" public that is ignorant, possibly misled, perhaps even irrational.

Of course, sometimes these factors—low levels of knowledge, susceptibility to anti-scientific arguments, even irrationality—are indeed actually present (see Sturgis & Allum, 2004), but this is not routinely the central explanation of differences of opinion between scientists and others. Science communication scholars have—for a quarter century and more—critiqued the "deficit" assumption on factual, theoretical, and practical grounds. Audiences, for example, are always mixed and often have substantial and relevant life experience to contribute to discussions of science and science policy. In the more contemporary view, communication should not be seen as a one-way process of transmitting "correct" information, but rather as a dialogic or two-way interaction between communicator and audience. A communicator who adopts a one-sided approach risks being seen as an arrogant "know-it-all" who does not understand how to listen—hardly an effective or ethical approach to fostering public trust in science. And public trust is much more important than knowledge as a factor in public acceptance of new scientific information.

Yet the assumption that improved knowledge transmission is the necessary "cure" for the "problem" of public acceptance keeps reappearing. Survey work continues to suggest that scientists assume the public's low knowledge of scientific issues necessarily shapes their perceptions of risk and of policy choices (Besley & Nisbet, 2013). Brossard and Lewenstein (2010) found that most outreach programs, even those that professed to be based on alternative models, still had deficit assumptions at their core. As Wynne (2006) noted for the UK, even national leaders who pay lip service to the ideas of participation and dialogue seem to continually reinvent the deficit model in response to each new crisis of public confidence.

This ongoing struggle with deficit model assumptions, an old story to science communication scholars, still deserves our attention. On the one hand, science communication does need to be rooted in more sophisticated conceptions of the ways publics can engage with science (that is, beyond listening to a lecture or reading a book). Focusing exclusively on the need for information transfer can, in practice, serve to absolve scientists and science communicators of a responsibility to acknowledge diverse values and legitimate doubts about the impacts of scientific advances. On the other hand, information is not irrelevant: Non-experts often do need expert help to make informed decisions on personal and policy matters, and science literacy does play a role in generating positive attitudes toward science. (Further discussion addressing some of these issues can be found in chapter 3 by Priest.)

Beyond the Science of Science Communication

The last few years have seen a proliferation of resources to help researchers develop their public science communication skills, due in part to national calls describing the need and in part to serving the increased ethic of engagement young researchers are bringing to their careers. And the National Academy of Sciences has promoted an idea called the "Science of Science Communication" through successive annual Sackler Colloquia (Fischhoff & Scheufele, 2013, 2014). These emphasized the need to move from intuitive notions of how science communication works to a research-based inquiry where "the communication of science is held to the same evidentiary standards as the science being communicated" (Fischhoff & Scheufele, 2013, p. 14032). While many scholars of science communication have subscribed to this idea for decades, this recent activity seeks to further legitimize science communication as an important topic of empirical study within the larger scientific community.

However, there is a parallel need to apply scholarly rigor to the analysis of the ethical questions faced by public communicators of science—a focus that unfortunately remains virtually ignored. This absence is not entirely surprising; ethical challenges that researchers encounter in the course of public engagement have also been neglected within the broader field of research ethics. A survey of the literature concluded that despite increased attention being paid to the field, associated ethical issues have hardly been considered (Meyer & Sandoe, 2012; see also Hollander, 2011). With a few exceptions, there has been little scholarship on the issues of communication ethics that confront scientists who communicate with broader publics. Yet within the broader communication discipline, it is widely understood that even very practical communication activity has important ethical implications (Holmes, Henrich, Hancock, & Lestou, 2009).

One aspect of this general neglect is actually rooted, in part, in disciplinary methodological preferences. While social and behavioral scientists can study some aspects of beliefs and value systems, including perspectives on ethics, by their usual empirical methods (surveys and experiments), others require more qualitative or analytical approaches. Yet it is precisely the emphasis on largely quantitative, systematic, empirical methods that often seems to qualify social scientific results as being, indeed, "scientific"—whether they are part of the "science of science communication" or concern other matters entirely. And more journals seem available in which to publish this kind of work. To better understand the ethical challenges facing science communicators, we need to broaden our approach to also include the methods more common in humanities disciplines (philosophy, history, rhetorical studies) and the more qualitative methods used in social science (interviews, observations, case studies). That is a big part of what we have tried to do with this volume.

Responsible communicators not only must master a toolkit of techniques, they also must be able to negotiate the often-conflicting communicative ideals and obligations inherent in complex communication situations. The conflicts researchers encounter when they leave their labs are many. How can researchers adapt to audience needs and interests without compromising accuracy and becoming salespeople instead of scientists? How can researchers contribute to often-heated civic deliberations without creating the impression that they are contributing to the politicization of science? How can they address those who may deeply disagree without appearing to adopt an authoritarian stance? How can they share the promise of scientific results without over-promising and arousing unjustified expectations? These and other ethical challenges arise in the ordinary course of communicating science. (Chapter 1 by Goodwin considers some of these in more detail.)

We hope this volume will serve to expand science communication scholarship from its present focus on science communication effects to broader questions of science communication ethics. Doing so will not only broaden our understanding of the larger public science communication environment, but also better prepare researchers for the difficult choices they are likely to face when engaging with the public.

A Note on the Emergence of this Project

The chapters in this volume came together as a result of several initiatives. In 2011, the first in a series of four summer symposia held at Iowa State University took place in Ames and was organized by ISU faculty members Jean Goodwin and Michael F. Dahlstrom. The series was designed to "jumpstart" discussions of science communication ethics. In part encouraged by that activity, the journal *Science Communication: Linking Theory and Practice* issued a call for contributions for a special issue on the topic of ethics and science communication, a joint effort between Susanna Priest, the journal's editor-in-chief, and Goodwin, who served as the issue's guest editor. Earlier work underlying three of our current chapters can be found in the resulting theme issue from 2012 (volume 34, issue 5). This resulted in discussions of producing a book and an open call for chapter contributions was issued in 2013.

Nine of the chapters included here were preceded by conference presentations at either the 2013 ("Ethical Issues in Science Communication: A Theory-Based Approach") or 2014 ("Normative Aspects of Science Communication") workshops in the ISU series, which provided these authors with opportunities to refine their ideas. Finally, two additional papers included in this volume were invited in order to fill specific gaps not addressed by the other material. Goodwin and Dahlstrom wish to thank the Iowa State University's Center for Excellence in the Arts and Humanities for their generous support of the workshops.

References

Besley, J. C., & Nisbet, M. (2013). How scientists view the public, the media and the political process. *Public Understanding of Science, 22,* 644–659.

Brossard, D., & Lewenstein, B. V. (2010). A critical appraisal of models of public understanding of science. In L. Kahlor & P. A. Stout (Eds.), *Communicating science: New agendas in communication* (pp. 11–39). New York, NY: Routledge.

Fischhoff, B., & Scheufele, D. A. (2013). The science of science communication. Introduction. *Proceedings of the National Academy of Sciences of the United States of America, 110,* 14031–14032.

Fischhoff, B., & Scheufele, D. A. (2014). The science of science communication. Introduction. *Proceedings of the National Academy of Sciences of the United States of America, 111,* 13583–13584.

Hollander, R. (2011, November). *Communicating and research ethics.* Paper presented at the Congress on Teaching Social and Ethical Implications of Research, Tempe, AZ.

Holmes, B. J., Henrich, N., Hancock, S., & Lestou, V. (2009). Communicating with the public during health crises: Experts' experiences and opinions. *Journal of Risk Research 12*(6), 793–807.

Meyer, G., & Sandoe, P. (2012). Going public: Good scientific conduct. *Science and Engineering Ethics 18*(2), 173–197.

Sturgis, P., & Allum, N. (2004). Science in society: Re-evaluating the deficit model of public attitudes. *Public Understanding of Science, 13*(1), 55–74.

Wynne, B. (2006). Public engagement as a means of restoring public trust in science: Hitting the notes, but missing the music? *Public Health Genomics, 9*(3), 211–220.

PART ONE

How Ethics Matters

Why should science communicators reflect upon ethics? Translating complex science for non-expert audiences may be technically challenging, but it does not seem like there are special ethical requirements involved, beyond the straightforward ones of honesty and accuracy.

This first set of chapters demonstrates how this view of science communication is not only naive, but can impede the successful communication of science by failing to meet the normative expectations held by audiences and the broader responsibilities of scientists within society. Ethical considerations are not a supplementary component to the practice of science communication—they permeate every decision in preparing for a communicative task, constructing a message, and interacting with an audience. Paying no attention to underlying ethical considerations does not necessarily make a science communicator act in an unethical way, however. Science communicators, like other human beings, may orient themselves to unstated and even unconscious ethical norms without deliberate thought. However, lack of conscious attention to ethics does tend to render them unprepared to navigate the complex social and cognitive environment faced when trying to communicate science to non-experts.

Later chapters will explore specific professional contexts and case studies that further illustrate these ethical considerations. This first part provides the system-wide view of how ethics underpins all of science communication. In other words, ethics matters.

The opening chapter by Goodwin (chapter 1) uses speech act theory to demonstrate how the act of communication itself is more than a mere transmission of ideas but instead a social contract involving ethical expectations and promises to fulfill them. Goodwin reminds us that audiences are active participants in the communication process who have no obligation to pay attention to messages and often have very good reasons for critical scrutiny of the messages to which they choose to attend. To overcome this justified skepticism, communicators need to communicate in ways that make them vulnerable to the ethical expectations of their audience. Goodwin unpacks these often contrasting ethical expectations for four roles that science communicators are often asked to fill: exercising authority, reporting, advising, and advocating. In sum, successful science communication depends on the audience recognizing that they are in a trustworthy and ethical relationship with the communicator.

The communication of risk is a fundamental communication challenge that arises within most scientific fields. Yet Thompson (chapter 2) demonstrates that the ways communicators and audiences conceptualize risk arise from distinct, yet often completely unnoticed, ethical frameworks. As a consequence, risk communication created with the best of intentions can provoke a range of unintended responses, including confusion, frustration, or even controversy. Thompson clarifies how approaches to risk are often based on either a utilitarian ethical framework—the ends justify the means—or a Kantian ethical framework—the ends never justify limiting the autonomy of individuals. Likewise, while experts often think of risk assessments as value neutral, the common use of the word "risk" entails value-laden questions of attention, loss, and necessary action. Thompson argues that while there is no one correct ethical foundation from which to build a conceptualization of risk, remaining oblivious to ethical choices will continue to frustrate the goals of risk communicators.

Why engage in science communication at all? Is the goal to influence audiences toward a more pro-science perspective or to provide information needed for informed personal and collective decision making? While both may appear reasonable, the first goal is based on strategically persuading audiences while the second puts more stress on transparent and unbiased content. Priest (chapter 3) explores this deep tension within science communication—the serving of self-interested strategic goals versus deliberative democratic goals. Because both goals are ubiquitous in the current media environment, Priest also argues for a need to equip audiences with enough understanding of science to distinguish when they are faced with potentially misleading strategic versus informative science content—a concept she calls critical science literacy. Priest applies these concepts to a climate change context to show the ethics may not be as clear-cut as they seem in the messy intersection of science and society.

Finally, Sprain closes this part with a discussion (chapter 4) that expands this seeming dichotomy between strategic and democratic science communication goals into the unavoidable ethical question of framing. Framing represents the focus through which a larger phenomenon is presented in a message. Sprain describes how an earlier call for scientists to frame their messages met with resistance when many scientists interpreted it as a recommendation for them to engage in inappropriate "spin" or manipulation. However, every instance of communication is framed; communicators must select what to include and what to omit within a finite message as well as the specific words with which to capture the intended meaning. Because framing is inevitable, science communicators must constantly make ethical choices about how to do it. To help navigate these choices, Sprain introduces the concepts of framing-for-persuasion versus framing-for-deliberations. While neither type of frame is inherently unethical, Sprain argues that over the long haul, framing-for-persuasion is likely to undermine the democratic deliberation necessary to incorporate science into the decision-making process.

The role of these early chapters is to provide broad conceptual frameworks that illustrate particular discipline-specific approaches to issues of science communication ethics. While these frameworks can, in principle, be applied to the topics in any of the subsequent chapters, it is not our goal here to attempt the construction of a single inclusive framework for analyzing science communication ethics. This book represents a very young interdisciplinary subfield. Our goal here is to stimulate a broader conversation about what the study of science communication ethics might look like; it would be quite premature to represent that conversation as completed.

1

Effective Because Ethical:
Speech Act Theory as a Framework
for Scientists' Communication

JEAN GOODWIN

Whenever we open our mouths, pick up a pen, or flip up the laptop to start typing, we draw on assumptions about how communication works. These assumptions guide us as we try to figure out what we are going to say and how we are going to say it. Although they usually aren't made explicit, worked out systematically, or grounded in evidence beyond personal experience, these assumptions could be thought of as our personal "models" or "theories" of communication.

The English language has one such theory built into its very vocabulary. Consider phrases like *put* ideas *into* words; *pack* more thought *into* fewer words; thought *content*; *get* an idea *across*; *convey* ideas; make sure the meaning *comes through*; *get* meaning *out of* words; *extract* meaning. These and other ordinary ways of talking about communication reflect what has been called the "conduit" (Reddy, 1979) or "transmission" model of communication. This model invites us to think of communication as primarily information transfer. A communicator is imagined to be packing an idea-object into a suitable linguistic container and sending the package through an appropriate medium (conduit) to the receiver, who then is supposed to unpack the container and add the idea-object to his mental store. This view has been elaborated in explicit communication theory, most famously in Shannon and Weaver's early transmission model (see Shannon & Weaver, 1949). But it is more widespread as an implicit mental model or folk theory. In particular, the conduit model of communication in general becomes the infamous *deficit model* when applied to the communication of science: the constantly reemerging, generally unstated assumption that the main task facing a scientist communicating with a nonexpert audience is to put her knowledge into understandable and interesting words in order to fill an audience's mental void.

As noted in the introduction, the failings of the deficit model of science communication are many and long understood. Of particular interest to readers of this volume, the deficit model follows its parent conduit model in being ethically impoverished. Both models imply that the communicator has only the limited ethical responsibility of making sure that the transmission system works. She is responsible for choosing packaging that will make it through the conduit and be frustration-free for the audience to open. We might summarize this by saying that under conduit/deficit assumptions, a scientist is communicating *ethically* if she is communicating *effectively*. Beyond this core duty to be an effective transmitter, she has no particular responsibilities to receivers, who are imagined in this approach to play only a passive role.

But is this all a scientist is responsible for? Wynne's classic case study (1989) showed how real audiences react when they are treated as empty-headed unpackers. Scientists who communicated about the dangers of Chernobyl fallout in the Lakes region of Great Britain succeeded mostly in generating distrust among affected sheep farmers because of failures of respect. The scientists didn't try to find out what the farmers knew about the locale, didn't consult farmers about how plans would impact their interests and practices, and didn't consider the troubled history of prior interactions between scientists and farmers. "Distrust" and "disrespect": these are ethically freighted words. Their presence reveals that there are ethical dimensions to science communication that go beyond the assumptions of the deficit model.

Many have called for replacing communication based on deficit model assumptions with approaches that emphasize dialogue or engagement between scientists and nonexpert audiences. While the ideals expressed in such calls are legitimate, one difficulty is that "dialogue" and "engagement" do not provide alternatives to the straightforward, intuitively compelling vision that the conduit model offers. "Dialoguing" just isn't as vivid as packing and unpacking, and ordinary English doesn't give us many metaphors for "engagement." So even scientists and science communication professionals who aim higher tend to find deficit model assumptions sneaking back unrecognized into their theories and practices (Brossard & Lewenstein, 2010; Davies, 2008; Wynne, 2006).

In order to stop reinventing the deficit model and to realize the promise of engagement, we need to cultivate a different set of assumptions—a different folk theory of communication. This alternative viewpoint needs to respect the audience's active role in communication; needs to allow for the development of a richer communication ethics; needs to fulfill the hopes expressed in talk of dialogue and engagement; and needs to be grounded in ordinary intuitions about how communication works. In the remainder of this chapter, I sketch how a perhaps surprising candidate—the philosophical theory of speech acts—can fulfill these four goals. A "speech act" is communication that in itself accomplishes a specific action, such as making a promise or lodging a complaint. As we will see, thinking of science communication as taking place through speech acts encourages us to pay attention to the ethics of communication.

In the following section, I summarize an interdisciplinary body of scholarship that draws inspiration from philosopher Paul Grice's original work, and show how speech act theory provides a general conception of communication that emphasizes the communicator's ethical responsibilities toward an active audience. I next demonstrate the power of speech act theory by using it to provide accounts of four quite different speech acts of special relevance to scientists contributing to public discussions: *exercising authority, reporting, advising,* and *advocating.* I close with a brief summary of the progress made, and a comparison of the speech act approach with the deficit model. In particular, I will wrap up by justifying my title: in the reverse of the view implicit in the deficit model, in the speech act approach science communication is only *effective* because it is *ethical.*

Speech Act Theory

To develop an alternative conception of communication, a good place to start is by considering what would happen if we allotted the receiver in the conduit model—that is, the audience—a more active role in the communication process. The conduit model assigns the audience the task of unpacking the message and adding it to their mental storehouse. But now let's give audience members some power: let's assume they are *autonomous* (they think and decide for themselves, relatively independent of outside influences) *agents* (they can make choices and do things that affect the world). Autonomous agency in the conduit model will become most apparent when the audience acts "badly" and interrupts the smooth transmission of information. For example: perhaps the audience refuses delivery of the message, or, on receipt, immediately tosses it in the trash.

These are not unusual ways to treat incoming messages. We do in fact dump junk mail in the basket right by the front door, fast forward through commercials, pay no attention to the flyers on the wall, and so on. Even when we invest attention, we often reject what salespeople, politicians, and pundits say. We even resist messages from those who have our interests at heart—our doctors, family members, or colleagues—when we don't like what they are telling us.

The conduit model invites us to take the transmission mechanism as just a given, with the main job of communication to be clearing away obstructions to it. To develop a better model of how communication manages to work, even facing an audience of autonomous and perhaps recalcitrant agents, we need to strip away this "given" status. We need to ask: what reasons does an audience have for paying attention at all, and if they pay attention, for trusting the quality of the message someone else is pushing on them?

Speech act theory offers an answer to these questions. In the approach to speech act theory developed by Grice (1957, 1969), extended by Stampe (1967), and applied to public discourse by Kauffeld (1998, 2001a, 2002, 2003, 2009), communication is fundamentally an *act*: something a communicator does with a reasonable expectation that it will change the world in some way.[1] Although we can use communication artifacts to change the physical world—like when we stack books to prop up a shelf—in general, communication does its work by affecting other people. These other people are presumed to be just as much agents as the communicator herself. They will only be willing to be affected by a message if they have good reasons to do so. Thus a key task facing any communicator is to provide her audience such good reasons for accepting her communication—that is, for trusting what she says. And the main way she accomplishes this is by openly taking responsibility for the quality of her message. Here is how it works.

Consider the very basic speech act of *saying* something. A communicator puts forward a sentence to an audience—for example, "Glucosamine pills help with the pain of knee arthritis."[2] We assume that it's not given that the audience will trust her. After all, believing is risky for the audience: the communicator could be mistaken or even lying (maybe she's a shill for a supplement manufacturer). Perhaps the audience has already heard of a study that found glucosamine no better than a placebo. Can they be confident that the communicator has put in the time and effort to verify her statement? Is she

1. Speech act theory originated in the work of Austin (1962), whose compelling argument that utterances do much more than just assert information opened the new field. For general overviews, see Green (2014) and Kauffeld (2001b).

2. The term "speech act" suggests that the communication is oral and my examples may have an oral flavor—I am a teacher of public speaking, after all. But in fact speech act theory aims to cover communication in any medium: oral, written, electronic, visual, and those yet to be invented.

lazy, negligent, or careless? In sum, the audience has plenty of good reasons to distrust what the communicator is saying—to resist it, to be recalcitrant.

Notice, however, that it's not just audience that may be getting into trouble. The communicator is also running a risk. In general, people are responsible for what they do intentionally. When she says something about the health benefits of glucosamine, the communicator intends the audience to believe it. She has thus made herself responsible for that belief; she is blameworthy if that belief is of poor quality, i.e., false. Furthermore, she's soliciting the audience's belief *openly*. If it turns out that glucosamine is ineffective, she won't be able to avoid criticism for lying or negligence; because she was open about her intention, she can't really wriggle out with excuses like "I didn't really mean it—I was just speaking offhand, mentioning something I read online somewhere—I didn't expect you to take me seriously." The fact that she is openly seeking belief thus ensures that the audience will be able to hold her responsible if what she says turns out to be wrong.[3]

Communication thus begins to look like a lose-lose situation: both communicator and audience are running a risk that the exchange will go badly wrong. The audience risks believing something false; the communicator risks being responsible for that false belief. But the communicator's vulnerability to criticism actually opens a way out of the apparent deadlock. The audience can reason as follows: (1) The communicator knows she is running a risk when she openly tries to get us to believe what she says. If her information about glucosamine turns out to be false, we can now hold her responsible. (2) She's a reasonable person; we can presume that she wouldn't put her good name at risk like this—with us, or even with herself—unless she was confident that the risk she faces is low. She must know that that she isn't lying and

---

3. Those who want to trace speech act theory back to its philosophical roots will find that the summary is an ordinary language rephrasing of the analysis developed by Grice, viz. S(peaker) will have said (and meant) that p, only if S produced an utterance U to A(uditor) with the following complex intention:

(I1) S intends that A respond (R) that p (e.g., they believe that p).
(I2) S intends that A recognize I1.
(I3) S intends that A recognize I2.
(I4) S intends that A's recognition of I1 and I2 provide A with at least part of A's reason for Ring that p.

As Stampe (1967) explains, in I2, S takes responsibility for inducing A to R. In I3, S takes this responsibility *openly*, thereby putting herself in a position not just to *be* responsible, but to be *held* so. This ability to hold responsible is what gives A a reason for his response, i.e., justifies I4.

hasn't cut corners in figuring out what glucosamine will do. (3) So if she's confident that the risk she faces is low, we can be confident that the risk is low for us as well.

The responsibility for the truth that the communicator has undertaken in saying something to the audience makes her vulnerable, and her open acceptance of that vulnerability gives her audience a good reason to trust what she is saying. To put it even more simply, in saying something seriously a communicator is not just transmitting information; she is transmitting information together with a personal guarantee of its truth. Her guarantee serves to alter the "social and moral order" (Kauffeld, 2001b); it changes the world just as effectively as a physical act like a hug. It creates, or at least enhances, a relationship between communicator and audience—a relationship in which the communicator is now responsible for speaking the truth, and the audience has a good reason to count on her to do so.

Saying is only the most basic among a very large set of speech acts. As we will see in more detail in the following section, a communicator's simple guarantee may not provide her audience with a good enough reason to satisfy their legitimate distrust. In such cases, the communicator will need to take on additional responsibilities to meet their additional concerns. Each such package of responsibilities constitutes a distinct speech act. Ordinary English has names for many: *promising, accusing, requesting, complaining, proposing, apologizing, commanding, thanking, warning,* and *challenging,* to mention only a few of the nearly one thousand verbs referring to things we can do with words (Verschueren, 1985). Despite this variety, the overall moral of speech act theory is simple: whenever you open your mouth to speak, you are opening a hole perfectly sized to fit a particular foot. Your audience can trust you to make sure that that foot gets nowhere near it.

It should be apparent that the fundamental view of communication put forward in this approach corrects two of the unfortunate features of the conduit model. Speech act theory positions audience members as autonomous agents in the communication process. It legitimates as reasonable their unwillingness to be passive recipients of information that's sent their way. All communication must be designed to respect the audience's right to think and decide for themselves, by providing them good reasons to trust what is being conveyed. Speech act theory also shows that communication effectiveness is grounded in communication ethics. In performing a speech act, a communicator undertakes responsibilities to her audience. She thereby sets up ethical standards for herself that she then has to meet, or else face criticism. It is precisely her vulnerability to ethical criticism that gives her audience good reason to trust her. Thus where the conduit model tells communicators to wrap their messages up in durable and attractive packaging in order to make sure they slide through the conduit with ease, speech act theory tells communicators to establish ethically sound relationships with audiences in order to make sure their messages deserve their audiences' trust.

If speech act theory provides an account of communication in general that is better than the conduit model, there is hope that it will also help us understand the communication of science in particular better than the deficit model does. In the following section, I take up four speech acts that have historically been of special relevance to scientists who want to contribute their knowledge to the public, especially in potentially controversial decision-making contexts: the acts of *exercising authority, reporting, advising,* and *advocating.* Table 1.1 offers a preliminary road map.

For each speech act, I draw from previous work on communication in policy controversies to clarify the ordinary situation that calls for the particular speech act and the reasons audiences have for distrust. I then explain the specific responsibility the scientist-communicator must undertake to create a good reason for trust, and sketch briefly how she can design her communication to live up to that responsibility, illustrated with an example. Finally, for each I note ways in which the transaction between communicator and audience can be extended beyond a one-time speech act to con-

TABLE 1.1. Some Speech Acts Characteristic of Scientist/Public Communication

Typical Situation	Audience's Reasonable Doubt	Communica-tor Undertakes Responsibility…	Speech Act	Some Typical Communication Features
The audience doesn't know something; the communicator does.	Is the communicator speaking within her field of expertise?	to speak as an expert—and only as an expert.	*exercising authority*	hedging: limitations, caveats, and uncertainties
The audience needs to understand a subject; the communicator is in a position to help.	Is the communicator adapting to our needs?	to empower the audience to assimilate the subject on their own.	*reporting*	a document designed to be explored and tested
The audience is making a decision but has a limited view; the communicator has a wider perspective.	Is the communicator meddling in our affairs?	to help the audience with their concerns.	*advising*	avoid telling; focus on audience's concerns, suppressing one's own
The audience needs to think through a matter; the communicator wants them to reach a particular conclusion.	Is the communicator trying to manipulate us?	to make the strongest possible case.	*advocating*	arguments, including counter-arguments against the always-present other side

stitute something more like an ongoing engagement between scientists and citizens.

Applications: Scientist Communicators and Decision-Making Audiences

EXERCISING AUTHORITY

The ordinary situation that leads to *exercising authority*[4] (Goodwin, 2001, 2010, 2011, 2015; Goodwin & Dahlstrom, 2013) is straightforward and ubiquitous. Non-experts need to know something. The scientist, an expert in her field, knows it. Therefore the scientist conveys her expert judgment, in order to fulfill the audience's need.

The central challenge of exercising authority is equally straightforward and ubiquitous: Audiences just don't like being subject to authority. The psychological theory of reactance suggests that people respond negatively when someone threatens to limit freedom of choice, with anger toward the source and counter-arguing against the message (Dillard & Shen, 2005). The more forceful the threat, the stronger the negative reaction. A scientist can exercise a compulsive force; when she announces her judgment, her non-expert audience really can't speak against it. So it is no surprise to find audiences recalcitrant in the face of messages from experts.

Still, it seems unreasonable for an audience to resist a scientist's judgment—especially when that very audience is paying to support the scientist as she gains and extends her expertise. But an audience can have legitimate concerns, over and above the natural tendency to resist being pushed around. The non-expert is in no position to assess the soundness of the expert's judgment. Or to be more exact, the non-expert *could* gain the capacity to assess expert judgment, but only by becoming an expert himself—something that's far too time-consuming to be worthwhile. This means that the non-expert is at the expert's mercy. The expert may be exercising an illegitimate authority by speaking outside of her field of expertise, either inadvertently or in a self-interested way. Unless the field difference is obvious (e.g., when chemist Linus Pauling promoted vitamin C as a cancer therapy), non-experts are unlikely to be able to catch the expert out. Critical thinking textbooks teach nascent citizens to question authority; they are right.

To meet these reasonable doubts, a communicator in exercising authority

---

4. That is, *epistemic* (knowledge-based) authority, in contrast to the authority police officers, judges, customs officials, etc., have to give commands; see Goodwin (1998).

undertakes responsibility not just to speak the truth, but also to speak the truth *as an expert*: in her expert role, within her field of expertise. She stakes her reputation as an expert on the soundness of what she is saying. By taking responsibility in this way, the expert opens herself to the risk of losing—or at least, undermining—her status as an expert if what she says turns out to be wrong. Her audience can presume that she would not rashly put her expert status at risk, and can conclude that she is being careful to stay within the bounds of her expertise. The expert's undertaking of responsibility to speak as an expert thus gives her audience reason to trust.

The basic attitude of an expert exercising her authority thus must be "with great power comes great responsibility." And furthermore, with great responsibility comes the potential for great trouble. The expert's sense of responsibility often shows up in her attempts to manage the trouble she may be getting herself into. The exercise of authority is perhaps paradoxically characterized by self-restraint, by phrases such as "no, it's more complicated . . . ," "yes, but on the other hand . . . ," or "we're not *sure*, although. . . ." These and similar caveats, limitations, and declarations of uncertainty are often thought to be carry-overs from scientific prose, unnecessary in public discourse. Instead, the account sketched here suggests that they are vital features of an exercise of authority. When the expert hedges, she shows that she is conscious of the heavy responsibility she is undertaking in speaking authoritatively, and is trying to limit that responsibility to just the matters she is most sure of.

Clear examples of exercising authority abound in health communication. The simultaneous growth of medical knowledge and of the internet has put many of us in the position of being avid yet reasonably skeptical seekers of authoritative information. Cochrane Reviews synthesize the best evidence on the effectiveness of interventions for expert audiences; their attached plain language summaries for non-experts are well designed to survive public skepticism. Both expert and non-expert audiences receive the same mix of judgments and caveats. A plain language summary starts by advancing authoritative statements like this recent one: "Water fluoridation is effective at reducing levels of tooth decay among children" (Iheozor-Ejiofor et al., 2015). But the summary carefully goes on to note that there was "insufficient information" about differential impacts of fluoridation on poorer and more affluent children, there were no studies of the effectiveness of fluoridation for adults, and the presence of methodological problems in virtually all the studies "makes it difficult to be confident of the size of the effects of water fluoridation on tooth decay." It might appear paradoxical: we trust the judgment *more* because its potential weaknesses are made apparent. But the Review's emphasis on its own limitations confirms our sense that the experts are aware of their responsibility to us, giving us good reason to trust them.[5]

The *exercise of authority* does not necessarily promote an ongoing dialogue between a scientist and her non-expert audience; she may simply announce her view, and then the transaction is done. But at times, one successful interaction with an expert can lead to another. An audience may at first trust an expert because she undertakes responsibility, and later come to trust her because they have through experience found her to be trustworthy, in what has been called the upwards "escalator of increasing trust" (Ensminger, 2001). Interactions begun with carefully managed responsibilities can lead to more open relationships, in which the expert may be able to drop the caveats without losing her audience's trust. Ideally, each of us will experience this with our healthcare professionals.

REPORTING

In the ordinary situation that leads to *reporting* (Kauffeld, 2012), a group (or individual) needs to understand some subject, likely because it is relevant to some judgment or decision they need to make. They aren't experts in the subject. They know they don't know, and thus are interested in being informed. That sounds relatively promising: unlike other audiences, the audience of a report is open to what is said. Reports are commonly solicited—audiences invite experts to draft reports, in a way that they may not invite advocacy or even advice.

Nevertheless there are challenges in making reporting work. The expert is being consulted because she is recognized as knowledgeable. But that very status can lead her audience to doubt whether listening to her will prove worth the effort. The audience may be legitimately concerned that the expert hasn't adapted her discourse to their needs. Perhaps the material provided will be too technical, or in an unfamiliar form, so they won't be able to un-

---

5. I am arguing that communicating uncertainties helps scientists meet the responsibilities they undertake when exercising authority. But will such hedges lead lay audiences to misinterpret the science being communicated? The empirical work on this question is limited and has shown mixed results. For example, there is some evidence that audiences do take advantage of poorly communicated uncertainties to "spin" results to support their preexisting beliefs (Dieckmann, Gregory, Peters, & Hartman, 2016). But at the same time, other work suggests that audiences are skeptical of strongly worded assertions, suspecting them of oversimplification (Winter, Krämer, Rösner, & Neubaum, 2015). Overall, empirical studies provide confidence that lay audiences are able to understand well-communicated scientific uncertainties (Fischhoff, 2012; Fischoff & Davis, 2014).

derstand it. Perhaps there will be too much material, which would waste their time—or too little, leaving them with only a partial view. Worse, the audience may fear that they won't even be able to tell whether the expert is being mis-, over-, or under-informative; not being experts themselves, they don't know enough about the subject to evaluate the accuracy or completeness of what she is saying. All of these are reasons for the audience to ignore, distrust, or even reject the material they themselves invited.

To meet these reasonable doubts, a communicator in reporting undertakes responsibility not just to provide accurate information, but also to do so in a way that will empower the audience to think the subject through for themselves. The reporter commits herself to presenting information that her audience will find accessible, that will put her audience in a position to appropriate the material for their own purposes, and even that will help her audience evaluate the report itself. In making a report, an expert undertakes responsibility to enable an audience to exercise their critical thinking skills, and allows—indeed invites—the audience to turn those skills back on her report. This undertaking gives her audience good reason to believe that the expert's report is adapted to their needs, thus giving them grounds to receive, consider, and trust it.

Reports are designed to fulfill this complex of responsibilities. A generic report has a "pyramidal" structure that allows the audience to go as much in depth as they care to—or as little. A report commonly opens with a brief summary of the most important points; the audience may find that this on its own is enough for their needs. If not, the body of the report develops the material at greater length, helping the audience assess the results by being open about the sources and methods used as well as potential limitations. A report also tends to have multiple roadmaps the audience can consult to find their way through the information: at least a detailed table of contents and cross-referencing between sections, and possibly an index, a glossary, or a bibliography of sources for still further exploration. Even if the report puts forward one conclusion as the best expert judgment of the reporters, considerations on all sides are raised and treated fairly, and sometimes dissenters are even allowed their own sections. To use the term suggested by McKaughan and Elliott (chapter 10, this volume), a good report will "backtrack," identifying all the values, interpretations, and frames relevant to the subject—both the ones the authors rely on as well as alternative viewpoints. All these textual features help the audience follow their own interests and assess for themselves the material being reported. Unlike an advocate's brief, which presses linearly toward one conclusion, a good report empowers each audience member to explore the subject in their own way.

The massive work products of the Intergovernmental Panel on Climate Change show how a good report affords self-directed assimilation and critical testing. It is unlikely that anyone would sit down to read the 1535 large-sized pages of the *Physical Sciences Basis* volume of the Fifth Assessment Report (IPCC, 2013). Instead, a reader is invited to start with the 29-page Summary for Policymakers (SPM). Those curious can go on to the 83-page Technical Summary, which "serves as a starting point for those readers who seek the full information on more specific topics covered by this assessment" by linking the conclusions of the SPM to specific sections of the fourteen chapters that make up the bulk of the report itself. Each chapter mirrors this organization, having its own table of contents, initial executive summary, and detailed discussion. The report facilitates critical evaluation by pointing out major uncertainties in the Technical Summary, by detailing limitations and uncertainties throughout, and by inviting frequently asked questions. Additional sources and data are available on the IPCC website for readers who want to test the report's conclusions in even more detail. Overall, the report acts as a sort of miniature universe, allowing non-expert readers to explore independently the evidence, theories, and controversies that make up our current understanding of climate change.

Although our attention in acts of reporting is often focused on the single, generally written, artifact that is issued—the report itself—in practice, the reporter's responsibility to the audience often plays out over an extended process. Since a central challenge of a report is that the audience doesn't quite know what it needs to know, the contents of the report are often the subject of an ongoing conversation between the reporter and the audience. In these conversations the goal or focus of the report is negotiated, helping both sides become more clear about what is wanted. The IPCC reports, for example, start with a series of meetings between experts and officials from participating countries to set the scope of the report, and they end with a similar series where the findings are approved, sometimes line by line. The ongoing conversation between reporters and audience can lead to a somewhat ironic result: Despite all the work that goes into them, many reports are never read, or at least not read beyond the executive summary. The process through which the report was created, more than the report itself, may be what helps the audience learn what they need to know. With that assurance, the audience can rely on the final product as an authoritative expert statement, and move on to carry on the affairs they wanted the report for.

ADVISING

In the ordinary situation that leads to *advising* (Kauffeld, 1999), a group (or individual) faces a decision about what to do. In general, a person is responsible for making up his own mind about matters that concern him; people hold themselves out to be autonomous, critical thinkers. Sometimes, however, a person lacks the breadth of perspective needed to make a decision wisely. An advisor hopes to remedy that situation. Often the audience knows that they lack perspective, and actively seeks out advice. At times, however, the lack of perspective extends to self-knowledge as well: The audience doesn't know that they don't know enough. In that case, the advisor may find herself thrusting advice on them.

In either case, the central challenge of advising comes from the fact that the advice is offered from outside. When an individual or group makes a decision, one of the most important features of that decision is that it is *theirs*: It is grounded in their concerns and is something they will have to live with. But advice is offered by *another*. The audience may have legitimate doubts about whether that other is taking their concerns into account. They may suspect the advice is really being offered for the advisor's own benefit. Or they may think that the advisor hasn't bothered to understand their point of view—that she arrogantly believes that she understands the audience's concerns better than they do themselves. In contrast to reporting, which is made difficult by the gap in understanding between reporter and audience, advising is made difficult by the potential for a gap in values between advisor and audience. Advice, including expert advice, can always be taken as intrusive meddling in another's affairs; even when the advice is good, it can deserve the response that "this is none of your business!"

To meet these reasonable doubts, in advising a communicator undertakes responsibility to help the audience determine what to do *about their own concerns*. She commits herself to addressing the decision not from her point of view, but from theirs. By taking responsibility in this way, the advisor opens herself to criticism if she fails to orient herself to her audience's perspective. Her audience can presume that she wouldn't rashly put herself thus at risk, and can conclude that she has given their concerns thoughtful consideration. The advisor's undertaking of responsibility to contribute to their concerns thus gives her audience reason to trust that she isn't just meddling, but is trying to contribute something that she reasonably believes the audience will find useful.

Undertaking the central responsibility of advising places significant limits on the advisor. An advisor will often find that there are many things that she can't say because they spring from her own perspective, not that of her audience. The experience may be familiar: A student comes in with two ideas for a project. One of them is conspicuously better than the other. But the teacher can't tell him that, because more important than the student choosing the *right topic* is the *student* choosing the right topic. It has to be *his* project. At best, the teacher can lay out the leading considerations (has the topic been done to death? can it be completed in the time available?) and nudge the student to see what is (to the teacher) obvious.

Agricultural extension services in the United States have developed successful advising mechanisms aimed at some of the world's most stubborn advisees: farmers. Extension personnel mediate between university-based researchers and the farmers who could benefit from innovations. They have learned from a century of experience that simply telling farmers what to do is ineffective, even when the answer is clear. Farmers are rightly convinced that they know their own operations—their own land and practices—better than any outside expert ever could. Extension advice is thus better offered through *showing* rather than *telling*, in events like field days where farmers can be given a guided tour of a new farming practice on an experimental plot managed by the university and/or another farmer. The field day not only conveys information, it opens opportunities for the farmer to express his concerns: to ask questions and point out differences between his operation and the one he is seeing. While the extension personnel design the framework for the event, at a good field day much of the communication is driven by the concerns the audience expresses. The audience, more than the advisor, is positioned as the agent.

Advice does not necessarily open the way to an ongoing conversation between the advisor and the audience. Indeed, it may be most appropriate for the advisor to offer the advice and then step back to leave room for the audience to make their own decisions. But at times, an "advising situation" demands a more extended engagement between communicator and audience. In order to credibly claim to be addressing the audience's own concerns, the communicator has to know what those concerns are. Thus a good advisor will often start by long and deep listening to the audience she hopes to serve. This is certainly the case for agricultural extension, where locally based personnel commit themselves to meeting the community's needs over the long haul.

ADVOCATING

In the ordinary situation that leads to *advocating* (Goodwin, 2012, 2013, 2014), the advocate sees that there is something wrong or missing in others' beliefs, attitudes, or decisions. She therefore undertakes to remedy that problem by persuading her audience to pay attention to something, think about it, change their beliefs or attitudes about it, and/or change their intentions toward it. In contrast to advising, which aims to help the audience make *their* own decision, in advocating the communicator aims to induce the audience to make the decision *she* knows is right (Pielke, 2007).

Here's the challenge: As before, audiences don't like being subjected to persuasion. When her audience detects that she is trying to influence them, they are likely to resist her attempts to limit their autonomy and restrict their freedom of choice. The fact that she thinks a decision is right doesn't bear much weight for others; they can legitimately distrust her motives in trying to bend them to her will. Any communicator who sets out to persuade is likely to encounter an audience concerned about being manipulated.

Communicators have two ways of meeting this challenge. The first is to try to hide advocacy under the cover of some less invasive undertaking. The advocate hopes that her advocacy may pass unnoticed, so audience reactance may not be triggered. One obvious problem with this approach is that it is often unethical: the communicator who tries to sneak "stealth advocacy" (Pielke, 2007) into her report (for example) will not be living up to her responsibilities as an advisor, since she will be arguing on the basis of *her* concerns, not those of her audience. Gaither and Sinclair (chapter 8, this volume) demonstrate similar ethical problems when corporate advocacy is hidden under cover of alleged reports on environmental topics. The sneaky approach is also simply unlikely to work. We have seen already that audiences are on the lookout for the covert attempts to persuade hidden under the cover of authority or advice. Especially in a culture like ours, saturated with attempts to sell products and persons, audiences are skilled in detecting and repelling attempts to meddle with their minds. If anything, audiences may be *over*-sensitive to persuasion, rejecting legitimate advice or appropriate reports because of suspicions of hidden manipulative intent. And if the audience does not call out the stealth advocacy, it is likely that advocates on the other side will. Over the long term, unethical covert advocacy is unlikely to persuade, but likely instead to get the advocate into trouble.

The prudent communicator will therefore pursue a non-sneaky approach, being open about her intent to persuade. But why should she expect her audience to be open to *being* persuaded? To understand how advocates can gain audiences, it's useful as usual to switch perspectives and consider why an audience might find advocacy worthwhile. As we have seen in the discussions of previous speech acts, audiences often know they need the benefit of others' knowledge and perspectives. It also happens that audiences realize they need to benefit from others' *reasoning*. When facing a difficult judgment, there are often many things to think about. But the reasoning process is time-consuming and laborious; sources of information have to be identified and assessed, a large amount of information has to be collected and analyzed, and the results need to be organized to identify lines of reasoning pro and con (or on multiple sides)—all this before the considerations are finally weighed and a judgment reached. Worse, this arduous process often takes place in an environment that is already crowded with competing communications, saturated with misinformation, partial truths, and contradictory reasoning. It would be very convenient if somehow a chunk of the cognitive labor involved could be outsourced to others who are competent to do it and motivated to do it well. If other people would take responsibility for sorting through the mess and synthesizing the best reasons for and against a particular judgment, then the audience would only have to review their work-product and make the final judgment: a still-challenging, but much less time-consuming, task.

The advocate can thus earn an audience's attention by accepting the job of "outsourced" reasoning: by taking responsibility for making the best possible case for her position. If she fails to develop strong arguments, she can be criticized not just for being ineffective, but for the more serious offense of failing to live up to her ethical responsibilities. The advocate's undertaking of responsibility thus provides her audience good reason to listen to the advocate and grant her a limited trust: to trust her to defend her position as zealously as she can.

Early rhetorical manuals, back in Cicero's Rome, recognized that the first rule of advocacy is to know what you are talking about. Thus scientists and other experts start off well positioned to become outstanding advocates. The practice of science is also quite argumentative, which should prepare scientists for the tasks of analyzing evidence, synthesizing reasons, and presenting the whole case clearly to the audience. Perhaps the most challenging feature of advocacy for a scientist will be finding that she will not be accepted as an authoritative voice. Advocacy is outsourced reasoning, but it is also openly partisan—an advocate presents reasoning for *one* position. Prudent audiences will thus always "hire" multiple advocates, who together generate the strongest reasons both pro and con (or on all sides). Of the full alphabet of possible considerations, if one advocate presents arguments A, B, and C, and her opponent X, Y, and Z, the audience can be reasonably assured that arguments M, N, and O aren't worth considering. But this means that a scientist-advocate may find herself in an uncomfortable position. Her arguments, likely based in her best scientific judgment, *will* be contradicted by another advocate—possibly an advocate without her excellent credentials. As an advocate, she cannot block even specious counter-arguments by simply saying "that's wrong—I'm the expert, I know"; she is obliged to keep on arguing.

Bill Nye has recently transitioned from science education to open advocacy for the theory of evolution, with a book, numerous public statements, and a much-watched debate with a prominent creationist. He is conspicuously fulfilling the advocate's basic responsibility: arguing enthusiastically and at length, to anyone who will listen.

Nye's campaign also illustrates another feature of advocacy: that it is often an ongoing process, requiring extended engagement with the audience. Changing minds takes time. The opposing side rarely capitulates quickly. So an advocate, even a scientist-advocate, should be pleased when she finds that there is occasion for more arguments to be made.

Conclusion

We see a similar pattern for each of these four speech acts, summarized in table 1.1 above. Communication generally starts when a public needs a scientist and the knowledge, understanding, perspective, or reasoning that she is uniquely equipped to provide (column 1). But the very asymmetry that makes what she says potentially valuable also makes it reasonable for the audience to be cautious in simply accepting it (column 2). The scientist can counter these reasons for distrust with good reasons for trust by openly undertaking a specific responsibility to the audience—by establishing local ethical norms that will guide their relationship (column 3). In undertaking just those responsibilities, the scientist is performing a distinct speech act (column 4). To fulfill the responsibilities she has undertaken, the scientist produces discourse that often has typical, generic features (column 5).

Speech act theory thus shows promise in going beyond the conduit metaphor/deficit model in meeting the four goals itemized in the introduction to this chapter. It provides an account of science communication that respects audience members' agency, giving them active roles in the communication process. It opens the way for a more complex communication ethics, identifying a variety of responsibilities that communicators undertake to meet the demands of different situations. It shows that extended interactions—ongoing dialogue and engagement between scientists and citizens—are often necessary to fulfill these responsibilities. And finally, speech act theory is grounded in ordinary intuitions about how communication works. While the explicit theory is a product of some sophisticated philosophizing (see footnotes 1 and 3), the speech acts that are the objects of the theory are everyday practices. We already know (at least in basic ways) how to exercise authority, report, advise, and advocate. We also already understand the responsibilities of each of these acts, or at least have a sense of the trouble we can get into as we undertake each one. Speech act theory does not replace our ordinary intuitions about communication; it merely articulates them more fully. This means that science communicators with no special expertise in communication theory already know enough to resist the blandishments of the conduit/deficit model. We just need to switch from asking "how" to asking "what." Whenever we catch ourselves wondering something like "how can I get this idea across—how can I put this complex topic into simpler words?" we can fend off such conduit model thinking and start by asking, "well, what am I trying to do here? Am I taking responsibility for being an advocate, or for giving advice?"

Of all the features of speech act theory, it is its ability to offer a more sophisticated ethics of science communication that makes it relevant to this volume. The deficit model assumes that science communication is ethical if it is effective: the communicator has fulfilled her responsibilities to her audience if she has delivered a well-packaged message. From a speech act perspective, science communication can be effective only if it is ethical: to earn her audience's trust, the communicator must undertake and then fulfill complex and situationally determined responsibilities to her audience. No ethical responsibilities—then no trust and no effective communication.

This aspect of situational determination also means that speech act theory shows promise of being flexibly adaptable to meet diverse communication needs. There is no single ethics of science communication; there are multiple responsibilities scientists can undertake and fulfill. The four acts presented here represent only a sample of approaches designed to deal with recurrent challenges of scientist/public communication. As different challenges arise, speech act theory provides a framework for identifying different ways to manage them. In fact, even these four speech acts can be combined in various ways: experts can issue *advisory reports*, for example, undertaking a dual set of responsibilities. It's worth noting, however, that not all combinations work as well. Some responsibilities are in tension with each other, and thus are difficult to undertake simultaneously. It may be challenging, for example, for a communicator both to insist on her authority and to function as an advocate; the former speech act admits of no reply, but the latter always contemplates the potential for counter-arguments.

I have been able to cover here only a few speech acts, with only a few practical pointers and examples about how they could be carried out; this essay is very far from a manual for scientists on "How to Do Science Communication Speech Acts." I have also focused exclusively on scientists' communication, not mentioning the communicative activities of others who play vital roles in the circulation of scientific information: media, traditional and new, government officials, politicians, advocates, opinion leaders, and ordinary citizens of all stripes. Still, it is the involvement of *scientists* somewhere in the process that makes science communication what it is. Finally, I have not had room to give accounts of the ways institutions can support—or undermine—the conditions needed to make scientists' speech acts possible. The presence of certain funding mechanisms, for example—governmental or industry— can foster or block trust by making it easier or more difficult for a scientist-advisor to credibly undertake a responsibility to be guided only by her audience's concerns. But in some sense, the institutions in which scientists and their audiences find themselves embedded are just additional contextual factors that have to be managed by speech acts on particular occasions.

The principles of ethical communication are not standards imposed from outside by institutions or other frameworks; these are not tablets of commandments passed down from on high. Instead, I have proposed a way of thinking about ethical responsibilities that shows them to be practically necessary to get communication to work before audiences who reasonably decline to be passive recipients of information. Communication based on a conduit/deficit model approach doesn't work because it doesn't confront the problem of legitimate distrust. As Wynne (2006) has commented,

> It is a contradiction in terms to instrumentalize a relationship which is supposed to be based on trust. It is simply not possible to expect the other in a relationship to trust oneself, if one's assumed objective is to manage and control the other's response. The only thing which one can expect to control, and to take responsibility for, is *one's own trustworthiness*. (pp. 219–220; emphasis in original)

By contrast, speech act theory provides the scientist a repertoire of mechanisms for making conspicuous what is generally the case: that she is worthy of trust. It asks her, though, to spend some time before preparing her effective communication to pause and think *what* she is doing.

References

Austin, J. L. (1962). *How to do things with words.* Oxford, UK: Clarendon Press.

Brossard, D., & Lewenstein, B. V. (2010). A critical appraisal of models of public understanding of science. In L. Kahlor & P. A. Stout (Eds.), *Communicating science: New agendas in communication* (pp. 11–39). New York, NY: Routledge.

Davies, S. R. (2008). Constructing communication: Talking to scientists about talking to the public. *Science Communication, 29,* 413–434.

Dieckmann, N. F., Gregory, R., Peters, E., & Hartman, R. (2016). Seeing what you want to see: How imprecise uncertainty ranges enhance motivated reasoning. *Risk Analysis.* doi:10.1111/risa.12639

Dillard, J. P., & Shen, L. (2005). On the nature of reactance and its role in persuasive health communication. *Communication Monographs, 72*(2), 144–168. doi:10.1080/03637750500111815

Ensminger, J. (2001). Reputations, trust, and the principal agent problem. In K. S. Cook (Ed.), *Trust in society* (pp. 185–201). New York, NY: Russell Sage Foundation.

Fischhoff, B. (2012). Communicating uncertainty: Fulfilling the duty to inform. *Issues in Science and Technology, 28*(4). Retrieved from http://issues.org/28–4/fischhoff/

Fischhoff, B., & Davis, A. L. (2014). Communicating scientific uncertainty. *Proceedings of the National Academy of Sciences, 111*(Suppl. 4), 13664–13671. doi:10.1073/pnas.1317504111

Gaither, B. M., & Sinclair, J. (2018). The ethics and boundaries of industry environmental campaigns. In S. Priest, J. Goodwin, & M. F. Dahlstrom (Eds.), *Ethics and practice in science communication* (pp. 155–174). Chicago, IL: University of Chicago Press.

Goodwin, J. (1998). Forms of authority and the real *ad verecundiam. Argumentation, 12,* 267–280.

Goodwin, J. (2001). Cicero's authority. *Philosophy & Rhetoric, 34,* 38–60.

Goodwin, J. (2010). Trust in experts as a principal-agent problem. In C. Reed & C. W. Tindale (Eds.), *Dialectics, dialogue and argumentation: An examination of Douglas Walton's theories of reasoning and argument* (pp. 133–143). London: College.

Goodwin, J. (2011). Accounting for the appeal to the authority of experts. *Argumentation, 25,* 285–296.

Goodwin, J. (2012). What is "responsible advocacy" in science? Good advice. In J. Goodwin (Ed.), *Between scientists & citizens: Proceedings of a conference at Iowa State University, 1–2 June, 2012* (pp. 151–161). Ames, IA: GPSSA.

Goodwin, J. (2013). Norms of advocacy. In D. Mohammed & M. Lewiński (Eds.), *Virtues of argumentation* (pp. 1–18). Windsor, Ontario, Canada: OSSA.

Goodwin, J. (2014). Conceptions of speech acts in the theory and practice of argumentation: A case study of a debate about advocating. *Studies in Logic, Grammar & Rhetoric, 36,* 79–98. doi:10.2478/slgr-2014–0003

Goodwin, J. (2015). Comment exercer une autorité experte? Un scientifique confronté aux Sceptiques [How to exercise expert authority: A case study of a scientist facing *The Sceptics*]. *Argumentation et Analyse du Discourse, 15.* Retrieved from https://aad.revues.org/2035

Goodwin, J., & Dahlstrom, M. F. (2013). Communication strategies for earning trust in climate change debates. *Wiley Interdisciplinary Reviews: Climate Change, 5*(1), 151–160. doi:10.1002/wcc.262

Green, M. S. (2014). Speech acts. In E. N. Zalta (Ed.), *The Stanford encyclopedia of philosophy.* Retrieved from https://plato.stanford.edu/entries/speech-acts/

Grice, H. P. (1957). Meaning. *Philosophical Review, 66,* 377–388.

Grice, H. P. (1969). Utterer's meaning and intention. *Philosophical Review, 78,* 147–177.

Iheozor-Ejiofor, Z., Worthington, H., Walsh, T., O'Malley, L., Clarkson, J., Macey, R., . . . Glenny, A. (2015). Water fluoridation for the prevention of dental caries. *Cochrane Database of Systematic Reviews* (6). doi:10.1002/14651858.CD010856.pub2

IPCC. (2013). Climate Change 2013: The Physical Science Basis. Contribution of Working Group I to the Fifth Assessment Report of the Intergovernmental Panel on Climate Change. Cambridge, UK: Cambridge University Press, 1535 pp.

Kauffeld, F. J. (1998). Presumptions and the distribution of argumentative burdens in acts of proposing and accusing. *Argumentation, 12*(2), 245–266.

Kauffeld, F. J. (1999). Arguments on the dialectical tier as structured by proposing and advising. In C. W. Tindale, H. V. Hansen, & E. Sveda (Eds.), *Argumentation at the century's turn: Proceedings of the Third OSSA Conference*. St. Catharines, Ontario, Canada: OSSA.

Kauffeld, F. J. (2001a). Argumentation, discourse, and the rationality underlying Grice's analysis of utterance-meaning. In E. T. Nemeth (Ed.), *Cognition in language use: Selected papers from the 7th International Pragmatics Conference* (pp. 149–163). Antwerp, Belgium: International Pragmatics Association.

Kauffeld, F. J. (2001b). Speech acts, utterances as. In T. O. Sloane (Ed.), *Encyclopedia of rhetoric*. Oxford, UK: Oxford University Press.

Kauffeld, F. J. (2002). Grice without the cooperative principle. In H. V. Hansen, C. W. Tindale, J. A. Blair, R. H. Johnson, & R. C. Pinto (Eds.), *Argumentation and its applications*. Windsor, Ontario, Canada: Ontario Society for the Study of Argumentation.

Kauffeld, F. J. (2003). The ordinary practice of presuming and presumption, with special attention to veracity and the burden of proof. In F. H. v. Eemeren, J. A. Blair, C. A. Willard, & F. Snoeck Henkemans (Eds.), *Anyone with a view: Theoretical contributions to the study of argumentation* (pp. 133–146). Dordrecht, the Netherlands: Kluwer.

Kauffeld, F. J. (2009). What are we learning about the arguers' probative obligations. In S. Jacobs (Ed.), *Concerning argument* (pp. 1–31). Washington, DC: National Communication Association.

Kauffeld, F. J. (2012). A pragmatic paradox inherent in expert reports addressed to lay citizens. In J. Goodwin (Ed.), *Between scientists & citizens: Proceedings of a conference at Iowa State University, June 1–2, 2012* (pp. 229–240). Ames, IA: GPSSA.

McKaughan, D. J., & Elliott, K. C. (2018). Just the facts or expert opinion? The backtracking approach to socially responsible science communication. In S. Priest, J. Goodwin, & M. F. Dahlstrom (Eds.), *Ethics and practice in science communication* (pp. 197–213). Chicago, IL: University of Chicago Press.

Pielke, R. A. (2007). *The honest broker: Making sense of science in policy and politics*. Cambridge, UK: Cambridge University Press.

Reddy, M. J. (1979). The conduit metaphor: A case of frame conflict in our language about language. In A. Ortony (Ed.), *Metaphor and thought* (pp. 164–201). Cambridge, UK: Cambridge University Press.

Shannon, C. E., & Weaver, W. (1949). *A mathematical model of communication*. Urbana: University of Illinois Press.

Stampe, D. (1967). *On the acoustic behavior of rational animals*. Photocopy. University of Wisconsin–Madison. Madison, WI.

Verschueren, J. (1985). *What people say they do with words: Prolegomena to an empirical-conceptual approach to linguistic action*. Norwood, NJ: Ablex.

Winter, S., Krämer, N. C., Rösner, L., & Neubaum, G. (2015). Don't keep it (too) simple. *Journal of Language and Social Psychology, 34*(3), 251–272. doi:10.1177/0261927X14555872

Wynne, B. (1989). Sheepfarming after Chernobyl: A case study in communicating scientific information. *Environment, 31*(2), 10–15, 33–39.

Wynne, B. (2006). Public engagement as a means of restoring public trust in science: Hitting the notes, but missing the music? *Public Health Genomics, 9*(3), 211–220.

2

Communicating Science-Based Information about Risk: How Ethics Can Help

PAUL B. THOMPSON

Risk science (or scientific risk assessment) is research conducted to ascertain *what* untoward or unwanted events might occur in connection with a given course of action (usually one involving a technical practice such as drug development or nuclear power), as well as the *likelihood* that these untoward and unwanted events will actually transpire, given specifiable contingencies. Risk communication is a contested activity conceived originally in terms of making the findings of risk science generally available to the public, but adjusted to include efforts for bringing public concerns and knowledge into the activity of risk science (Priest, 2009). The aim of this chapter is to survey two important points of intersection between the communication of scientific or science-based findings on risk and social ethics. The first concerns a conceptual bias common to most scientific risk assessments that leads communication efforts to emphasize one general class of ethical norms at the expense of others. The second involves a grammatical bias that puts scientific communication efforts at odds with common ways of speaking and writing about risk taking and risky situations. The chapter thus takes on a somewhat limited subset of the topics that a broader survey of ethical issues associated with risk assessment might encompass (see Cranor, 1990; Hansson, 2007).

Ethics: A Brief Clarification

Philosophers and other scholars who work in and on ethics are familiar with a number of ways in which their topic confuses and puzzles their audiences.

This chapter has been updated and modified from an article previously published in *Science Communication: Linking Theory and Practice*, 34(5): 618–641, October 2012.

Some of this mayhem can be traced to the very grammar of the word "ethics." Ethics (noun plural) are norms, standards, and expectations that circumscribe, direct, and possibly motivate conduct appropriate for a given situation. Among professional groups, ethics (p.) specifies rules or codes for practices thought to be essential for or peculiarly characteristic of the profession. They may be articulated by a list of standards that stipulate types of behavior for specific situations, or by principles that prohibit conduct deemed to be inconsistent with professional norms. For example, the code of ethics for professional appraisers forbids someone who makes an appraisal for a fee from offering to purchase the item being appraised. Ethics (p.) also indicates tenets or canons applicable in common life: "don't tell a lie," "always be courteous." Here, ethical expectations are frequently communicated through stories or the celebration of iconic individuals (heroes and villains) in a manner that does not easily translate into imperatives indicating specific actions. Journalistic ethics, for example, might be promulgated by rules such as "Always tell both sides of the story" or by reference to an individual (such as Edward R. Murrow or Katharine Graham) whose conduct was thought to exemplify good ethical character.

Ethics (noun singular) is a domain of discussion, discourse, and theory devoted to the critical analysis of both individual and social conduct. In academic circles it is sometimes characterized as a subfield of philosophy, though academic programs in ethics have tended to be interdisciplinary. Sociological studies in ethics, for example, often undertake empirical work to identify the norms, value judgments, and opinions about ethics (p.) that are most widely shared in a given social group. Philosophers and practitioners of ethics (s.) may be engaged in a critical debate whose purpose is to forge agreement on what actions should be undertaken in a given context, as distinct from those actions and practices that typically *are* undertaken. There may also be discussion, discourse, and theory devoted to more general structural, logical, and psychological dimensions of ethics (p.). *Ethical theory* is an attempt to derive a very general set of prescriptive procedures that identify right actions, while *metaethics* is an attempt to characterize the nature of morality and ethical conduct without necessarily offering any basis for prescriptive judgment.

However, ethics (p.) and ethics (s.) are not fully distinct. On the one hand, as already noted, key aspects of ethics (s.) involve the empirical study of ethics (p.). Attempts to formulate ethical theories or complete metaethical analyses are increasingly informed by these empirical studies (Appiah, 2008). On the other hand, there are often situations in which commonly accepted norms either conflict or do not fully specify the conduct that is demanded of people who wish to act in an ethical manner. In such situations, the form of critical inquiry typical of ethics (s.) may help the person or persons choose which of several courses of action are appropriate. Similarly, circumstances can arise in which large scale cultural, organizational or group change calls for widespread reexamination of traditions, habits, and norms. Transitions associated with changing views on gender, race, and sexuality have led to extensive ethical debates in recent years, and these debates are widely understood to have practical as well as scholarly significance.

The following discussion of ethics and scientific risk communication is more typical of ethics (s.) than ethics (p.). There is no attempt to specify rules or standards for the practice of risk communication. Rather, the goal is to expose how ethical assumptions penetrate deeply into the way that risk itself is conceptualized, shaping the formulation of risk analysis and risk management in particular situations. The thesis is that the practitioners in scientific risk assessment and communication efforts that emphasize scientific findings on risk have often foundered (sometimes badly) in ways that can be illuminated by ethics (s.). In some cases, they have implicitly made normatively biased framing assumptions while in others they have adopted conceptualizations of risk that are naively oblivious to ambiguities in the way that the word "risk" functions grammatically. Either way, my thesis presumes that speakers of English possess a broad competency for using the word "risk" in ordinary conversation. Speakers of English use the word "risk" along with related grammatical forms, such as the adjective "risky," the adjectival nominative "riskiness," or the gerund "risking," with the fluency characteristic of a native speaker. Native speakers are considered to have authoritative opinions about the meaning of words in their native language due to a natural acquisition process that cannot be matched by those who learn the language later in life. Native speakers will not necessarily be able to articulate grammatical rules for the language, but will have an intuitive understanding of grammar through their experience with the language (Love & Ansaldo, 2010). The concept of risk is circumscribed by the meanings that can be derived from this usage. As will be shown below, these grammatical forms associate the concept of risk with ethical content: They typically convey a value-orientation toward the conduct or events that are described as risks or as risky.

However, as will also be discussed below, the precise nature of this value orientation varies from one context to another. Such variance in ethical content is normally unproblematic in ordinary conversational usage, because context resolves the potential for ambiguity. People can generally follow a conversation about risk much more readily than they can follow the definition and distinction-drawing that are typical of work done by philosophers. What is true of people in general is often true of disciplinary specialists who have not been socialized into the peculiar obsession with consistency and linguistic precision that is typical of academically trained philosophers. There is thus some risk (no irony intended) that readers will find the analysis that follows pedantic, boring, and unengaged with the concerns of science communication. But what is true of philosophical discourse is also true of science-based risk communication. In contrast to the general concept of risk that unifies usage among contemporary speakers of English, a specific conception of risk is developed in scientific risk assessment, where a given methodology or problem solving task has been presupposed (Thompson & Dean, 1996). For example, financial analysts and epidemiologists are both highly conversant in risk problems, but it is clear that they conceptualize risk in very different ways. Not only are the outcomes (financial loss vs. epidemic disease) quite distinct, financial losses are incurred by the individual or corporate entity whose assets were "at risk," while the epidemiological conception of risk relates observed outcomes to a general population. Given their methodological orientation, some epidemiologists insist that risk is definable only in reference to populations, but taken literally such a statement implies that financial analysts who discuss the risk of a particular investment decision simply don't know what they are talking about!

It is uncontestable that specific conceptualization of risk are necessary for specific research, technical, decision, and analytic activities. It is also a commonplace of science communication that experts are advised to maintain cognizance of the way in which their specialized knowledge may not be shared by the larger public. Yet it is a thesis of this chapter that this commonplace is not fully appreciated with respect to expert conceptions of risk and the communication of science-based information about risk to the larger public.

Ethical Issues in Risk Assessment

The recent history of contested technologies exhibits a persistent tension between the perception or estimates of scientifically trained experts and the risks associated with these technologies by the lay public. In an early and frequently cited article Judith Bradbury draws a distinction between a technical community of risk experts who presume that risks are objective features of the world, and a second community of experts in the social and behavioral sciences who presume that risks emerge as a blend of perception, culture, and the social situation of parties who experience risk. Experts in the technical group presume that risk is fully described by two discrete quantities. The first is a scientific account of some phenomenon or state-of-affairs, and the second is the probability or likelihood that this phenomenon or state-of-affairs will occur. Experts in the second group are more likely to view variation in a given person's state of knowledge, their values about what matters, and more conventional social variables such as race, class, and gender as relevant to risk. Bradbury wrote that experts in the second group believe that risks are sociocultural constructions (Bradbury, 1989).

My own approach differs from Bradbury's "social construction" in two respects. First, as this chapter will discuss at some length, language conventions place boundaries on experts' ability to specify technical conceptions of risk that go well beyond the influences of race, class, and gender or of divergent values. This is not to imply that racial, class, and gender biases are absent from the way that either experts or lay publics conceptualize risk nor to deny that such sources of bias raise ethical issues. They are just not the issues that are the primary topic of this chapter. Second, the term "social construction" has plunged risk communication into a difficult ontological debate. On some readings, social constructions are free-floating cultural artifacts that are to be sharply distinguished from "real things" that are amenable to scientific research (see Hacking, 2001). By the end of the 1990s, risk scholars were writing about "the risk wars," implying that there were two antagonistic schools of thought founded on wholly distinct concepts of risk. Natural scientists held that risks are "real" and amenable to the methods of the biophysical sciences, while social scientists in the social constructionist school of thought were most responsible for bringing communication studies to the forefront, especially the so-called Orange Book. This National Research Council Report (entitled *Understanding Risk*) emphasized a number of ways in which lay publics faced challenges in understanding the quantifications of risk typical of the technical community. The report also took the technical community to task for neglecting legitimate concerns of lay publics that were not easily incorporated into the probabilistic approach to risk (Slovic, 1999).

This way of interpreting the field of risk studies would leave anyone attempting to address problems in risk communication in something of a quandary. Does one presume that technical experts have it right, making the job of risk communication one of translating technical results into more understandable language? Or does one follow a line of thought that suggests lay people can have important input into risk assessment, making risk communication into an activity of translating from lay publics to technical experts? The possibility that both roles are legitimate was obscured by the tendency to interpret the two schools of risk as arising from incompatible conceptual paradigms. Elsewhere I argue that while experts in biophysical and social sciences may indeed have distinct conceptions of risk, it would be erroneous to presume that non-expert members of the public are "contextualists" or "social constructionists" whose understanding of risk matches that of the social scientists, rather than the technical experts (Thompson, 1999).

The native speakers I will regard as authoritative sources on the meaning of risk would be as mystified by talk of social construction as they would be by the detailed statistical models of epidemiologists. The distinction I will emphasize below is between *any* expert conception of risk (including those based in the social and behavior sciences) and a general concept of risk that underwrites ordinary conversational uses of the word "risk."

Risk Science and Ethical Bias

Although philosophers have developed many approaches to the theorization of ethical conduct, two patterns of thought were especially influential in the twentieth century. Utilitarian ethical theories define right action as a function of the consequences or outcome produced or caused by it. Kantian, Neo-Kantian, or deontological ethical theories (henceforth simply Kantian) define right action according to its conformity with rules and without regard to outcomes. Although there are numerous difficulties in both approaches, my focus here concerns the way that risk is implicated within and addressed by each approach. Both assume that the function of an ethical theory is to dictate which of numerous possible actions is ethically correct. They are designed to address a choice situation in which a decision maker explicitly and deliberatively considers what should be done.

In definitive formulations of utilitarianism developed by Jeremy Bentham (1748–1832) and John Stuart Mill (1806–1873), ethical decision making is portrayed as a problem in evaluating alternative actions in terms of their expected benefit and harm. An ethical decision maker treats choice as an optimization problem where the worth of decision outcomes to all affected parties is a quantity for which an optimal value is sought. This very general specification leaves many interpretative questions open. What counts as a benefit or a harm? How should trade-offs between benefit to one party and harm to another party be reflected in the optimization process? Who counts as an affected party? For example, Bentham is often noted for his opinion that decisions should take account of their impact on non-human animals (Singer, 1975; Derrida, 2008). What is relevant in the present context is that ethical conduct is a function of the expected value that can be assigned to the consequences of each decision option, once these admittedly difficult interpretive issues have been resolved.

In contrast to the utilitarian approach, a Kantian stresses the way in which any decision maker will make choices according to some principle or rule. Immanuel Kant (1724–1804) provides an approach for testing decision rules according to a master rule, the categorical imperative (Kant, 1785/2002).

Decision makers are instructed to ask themselves whether the principle justifying their own action could serve as a universal law, as a principle that would be used by any decision maker to determine what is ethically correct for any and all relevantly similar cases. One point of this question is to make the decision maker aware of possible sources of favoritism or bias. In particular, asking whether one would be willing to have others make decisions according to a given rule should sensitize a decision maker to whether or not they would consider the decision to have been ethically correct even when they were in the position of an affected party. Hence some have suggested that the overall thrust of Kantian ethical theory is to promote fairness as a standard for ethical decision making that overrides all other principles (Rawls, 1980), while others have suggested that the categorical imperative is roughly equivalent to the Golden Rule: Do unto others as you would have them do unto you (Hirst, 1934).

Those who believe that science does not take sides on ethical issues will want to claim that scientific risk assessments are neutral bystanders in the philosophical debate between utilitarians and Kantians. But Bentham was clearly aware that the consequences of a decision can rarely be predicted with certainty. His approach built on the work of decision analysts who had studied gambling problems. The utilitarian response to uncertainty is to represent both benefit and harm as quantities of worth or value that reflect the probability that the beneficial or harmful outcomes will actually materialize (Bentham, 1948). Thus, Bentham's approach prefigures that of twentieth-century decision theorists who characterize decision making as a problem of making the best available trade-off between risk and benefit, and who presume that a risk analysis advises the decision maker of both the value and harm associated with potential hazards and also the likelihood that the hazards will actually occur. Much of present-day risk analysis operates under the implicit assumption that decision makers understand their task in the terms that Bentham specified more than 200 years ago. As a matter of intellectual history, at least, risk assessment is a direct intellectual descendent of utilitarian ethics.

However, this is not to say that Kantians have nothing to say about risk. Intuitively, application of the Golden Rule suggests that an agent should be particularly sensitive to the impact of his or her action on freedom of others. Indeed, if other parties remain free to act according to their own lights, one may presume that the decision under review has little or no ethical significance from a Kantian perspective. For many who take this approach, actions that affect only the agent himself are not strictly moral at all, but should be considered purely prudential (Vaccari, 2008). From the Kantian perspective, it is mostly situations in which risks are imposed upon others that ethical evaluation is required. From the perspective of Kantian ethical theory, real-life cases of risk-based decision making show that the utilitarian's risk-benefit approach is unethical. In the 1930s the US Public Health Service conducted extensive research on syphilis, including an observational study of untreated syphilis in Macon County, Alabama. The Tuskegee Study continued into the 1970s, well after penicillin had been recognized as an effective treatment for syphilis. The researchers' rationale for allowing the study to continue was influenced by multiple factors (Rothman, 1982), but the National Institutes of Health report on the incident (henceforth, "the Belmont report") stressed the way that a narrowly utilitarian approach to the ethical evaluation of medical research led researchers astray (Jones, 1993).

A utilitarian evaluating the Tuskegee Study might have reasoned that benefits from continued observation of patients suffering from the untreated disease would be justified by a better clinical understanding of the disease and its effects. (I do not assert that all utilitarians would reason this way, only that this pattern is discussed in the literature on Tuskegee and the Belmont report [see Jones, 1993].) For a utilitarian, these benefits could potentially offset the risk of harm associated with continuing to observe disease in the research subjects enrolled in the study. Nazi scientists also rationalized horrific research on prisoners being held in concentration camps, apparently applying a similar application of the utilitarian maxim. However, if there is any instance in which the test suggested by the categorical imperative would suggest that utilitarian reasoning cannot be universalized, the case of the Nazi doctors would qualify (Macklin, 1992). Following the Belmont report, the principle of informed consent has been the standard for ethical acceptability of risk to human subjects in research settings. Succinctly, research that imposes risk on human subjects is ethically acceptable only if subjects have been fully informed of all risks and researchers have secured freely given consent. From the perspective of Kantian ethical theory, the utilitarian emphasis on outcomes introduces bias because it directs a decision maker's attention away from the factors of greatest ethical relevance.

To summarize this section, the conceptualization of risk as an expected value that reflects both hazard and exposure is historically and conceptually tied to utilitarian approaches in ethical theory. Bentham would have agreed that the best way to understand risk is to evaluate the harm that would occur when a hazard materializes, and to reflect the decision maker's uncertainty about whether the hazard would materialize probabilistically. In contrast to this approach, Kantian ethical theory stresses the dignity and autonomy of affected parties, and views the imposition of risk upon any affected party as a potential affront to that person's dignity and autonomy. The relative value of either hazard *or* the probability of exposure do not figure prominently in deriving the ethical norms pertinent to risk imposition in the Kantian approach.

Ethical Bias in Risk Communication

One important strand of research on risk communication involves problems in which communicators hope to induce behavioral change among individuals and social groups in light of research on risk factors that contribute to mortality and morbidity (Fan & Holway, 1994; Yanovitzky & Stryker, 2000). The presumption behind these efforts is that target groups do not know that their behavior is risky, and that an effective information campaign will induce voluntary behavioral change once people have become aware of the relationship between their conduct and their exposure to hazards. Cates, Grimes, Ory, & Tyler (1977) described an apparently successful risk communication effort in the public media to advise women of the risks associated with leaving an IUD in place after conception. Conveying science-based information about the hazards of failing to remove an IUD led to rapid behavioral change among women and to the desired decline in injury to health for pregnant women and their unborn children. Risk communication efforts associated with smoking, HIV-AIDS, and numerous other behaviors have had mixed results in changing behavior.

The evaluation of these cases from the perspective of competing ethical theories not only illustrates the difference between the ethical principles endorsed by utilitarians and Kantians, but also shows that this difference may not matter much for some paradigmatic cases of risk communication. From a utilitarian perspective, the criterion for ethical management of risk is to achieve optimal trade-offs between beneficial and risky activities. Since the benefits from activities that expose people to the hazards of IUDs, smoking, and HIV-AIDS appear trivial in comparison to the harm that prevails when risk is realized, a reasonable application of the utilitarian maxim suggests that behavior change to recognize these risks is strongly indicated. If so, the standard for ethical success in risk communication would appear to reside in its impact on behavioral change. The principle of informed consent, in contrast, requires only that people actually know what the relevant risks are, and not that they also engage in behavior change. An Environmental Protection Agency official appeared to endorse this view when he wrote: "Success in risk communication is not to be measured by whether the public chooses the set of outcomes that minimizes risk as estimated by the experts. It is achieved instead when those outcomes are knowingly chosen by a well-informed public" (Russell, 1987, p. 21).

But cases such as IUD use and HIV-AIDS are also ones in which risk factors are associated with intentional actions that are under the respective individuals' control. On the one hand, if there are truly negligible benefits associated with a risky activity, as would seem to be the case in failing to remove an IUD, informed consent can be expected to produce the behavioral change indicated by the utilitarian maxim. This is therefore a case in which any reasonable person will almost certainly respond to a risk communication message with behavioral change. There is no reason to expect divergence between what individuals choose and what experts would regard as optimizing behavior. On the other hand, when there is little chance that interference in individual freedom will succeed in inducing a desired behavioral change, it is unlikely that a utilitarian using the optimizing approach will ever recommend actions that would violate the categorical imperative. In the case of sexual behavior, the long history of failed efforts to regulate behavior provides a reason not to recommend risk management strategies that would diverge from what is required by a principle of informed consent: that is, providing the information on risk and simply hoping for behavioral change. The theoretical difference between optimizing and informed consent makes little practical difference in either case.

However, there are risk policy domains where ethically based debates can rage. The management of risks from traffic accidents illustrates the point. From a utilitarian perspective, mandatory precautions such as seat-belt and motorcycle helmet laws are justifiable because they save lives (Hauser, 1974; Watson, Zador, & Wilks, 1981). From a perspective that stresses individual informed consent, such laws intrude upon the freedom of an individual decision maker (Irwin, 1987; Rollin, 2006). As with sexual activity, decisions about whether to use seat belts or helmets do not directly cause a hazard to anyone other than the decision maker (though this is not to deny that there may be social costs or indirect effects on third parties). From the standpoint of informed consent, it is reasonable to provide drivers with information, but there would be no ethical imperative to ensure that they use information in the manner that a risk optimization paradigm would suggest. Thus the assumption that risk communication is intended to induce behavioral change would appear to have an underlying commitment to utilitarian ethics.

When an action imposes risk on someone other than the decision maker, the informed consent considerations that arose in connection with the Belmont report become relevant. Clearly, many activities in contemporary society impose risk on others, yet provide little or no opportunity for giving or withholding consent to the host of other industrial activities. Decision making on the regulation of these activities falls to public authorities. Debate over this decision making mirrors classic philosophical debates over whether public policies should aim to optimize risk and benefit, as utilitarianism claims, or whether they should strive to protect affected parties from intrusion into their sphere of personal freedom. An extreme interpretation of the Kantian view might hold that there are no cases in which environmental risks could be ethically tolerated short of an effort that actually secures consent from affected parties (Machan, 1984). More realistically, the view might be that our political system reaches a compromise in which the willingness of most citizens to accept risks in exchange for the benefits of the economic activity derived from polluting activity reflects a kind of implied consent (Killingsworth & Palmer, 1992). In either case, the function of risk communication resides in the need for those who bear risks to have a clear understanding of them, rather than in whether or not their behavior produces optimal risk-benefit trade-offs.

In contrast to my claim that science-based risk communication needs to be attentive to this tension between utilitarian and Kantian ethics, a great deal of the scholarly literature on environmental risks appears to simply assume that the utilitarian viewpoint is ethically correct. Cross (1998) writes that what matters ethically is whether optimal trade-offs are made. Public opinion figures indirectly in the process of reaching this optimum, because it is widely believed that irrational attitudes, heuristics and biases, misperceptions and the NIMBY[1] syndrome can create political roadblocks to the implementation of those policies that would, in fact, most fully satisfy the optimizing goals of the utilitarian maxim (Starr & Whipple, 1980; Lewis, 1990). For a thoroughgoing utilitarian, the goal of risk communication is unabashedly one of manipulating opinion and behavior so that an optimal balance of risk and benefit can be achieved. This need not imply that the question of whether risks are voluntarily accepted is irrelevant to a utilitarian. Chancey Starr, one of the founding figures in contemporary risk analysis, argued that standards of risk acceptability for involuntary risks would indeed be much higher than for risks that people incur when they voluntarily engage in high risk activities. But Starr's approach was thoroughly utilitarian in its ethical commitments (Starr, 1972). It is also quite possible to reconcile utilitarianism with the need

---

1. NIMBY is an acronym for "not in my back yard." The NIMBY syndrome is the tendency for people to form risk attitudes and engage in political action when risky activities are geographically proximate, but to have different attitudes when risks are described abstractly or are borne by distant others. The NIMBY syndrome implies that people are active on risk issues only when their self-interest is involved.

for good faith communication efforts, particularly when non-experts have information that experts lack (Wynne, 1996). Knowing how people will act in a given situation will generally be important for anyone who hopes to achieve the greatest good for the greatest number. Placing people in a position where they can engage in politics on an informed basis may not be the most efficacious way to achieve an optimal ratio of benefit and risk.

Policy change over smoking represents an interesting and illustrative case. Before research results demonstrated the hazards to third parties from secondhand smoke, there had been little momentum behind efforts to introduce mandatory laws governing smoking behavior. Although there was a gradual increase in legislation intended to limit or discourage smoking from the early 1970s, during this era smoking was largely conceptualized as an individual behavior. Exposure to secondhand smoke was regarded as an annoyance by nonsmokers, but not as an activity that exposed them to risk (Syme & Alcalay, 1982). Once the health effects of exposure to secondhand smoke became well understood, a new rationale emerged for legal remedies (Walsh & Gordon, 1986; Ezra, 1990). In the ethical terminology developed here, it became possible to see smoking as a behavior that imposed risk on nonsmokers, and given that ethical framing of the issue, a morally compelling case for legal bans on smoking in public areas could be mounted by mobilizing rationales derived from Kantian ethical theory. Laws that regulate exposure to secondhand smoke have proliferated in the past quarter century (Eriksen & Cerak, 2008). While a consistent utilitarian would have had enough reason to support policies against smoking even in the absence of scientific findings on secondhand smoke, it was only when smoking could be seen as imposing a risk on others that change began to occur. Green and Gerken (1989) characterize this as a victory of self-interest over ethics, implying that there were no compelling ethical reasons for change to occur. But Kantians might well have viewed smoking as a purely voluntary risk prior to the emergence of scientific results on secondhand smoke, then realized that the rights of affected parties (themselves or others) were actually at stake. From a philosophical perspective Green and Gerken exhibit a strange blindness to the difference between utilitarian and Kantian thinking, illustrating that such blindness can be found among both social and biophysical scientists.

There have been other cases where the ethical framing for a risk issue has remained contentious. Labeling for food containing ingredients derived from genetically engineered crops (or so-called GMOs) provide an example. When these foods began to appear on US markets in the late 1990s, officials at the US Food and Drug Administration (FDA) determined that since there were no health risks associated with consuming these foods, no benefit could be derived from providing labels advising consumers of their presence. This policy was a fairly straightforward application of utilitarian reasoning and was defended as such in analyses that supported the FDA's decision (Vogt & Parrish, 1999). Arguments supporting labels called upon citizens' "right to know" and linked values that individual consumers might hold regarding personal risks with religious and other personal freedoms that are relevant to food choice (Thompson, 2002).

In summation, the overall ethical framework in which a given risk problem is conceptualized can play a significant role in shaping the way one would develop an appropriate communication effort. In cases where informed consent has been clearly identified as the appropriate standard, as in developing protocols for the use of human subjects in research contexts, a communication tool should clearly try to minimize its persuasive component. At the same time, some of the most widely studied problem areas involve educational and persuasion efforts where the attempt to achieve behavior change would appear to be ethically well justified. But the role of risk communication in a wide range of policy-relevant cases is far less clear. Is it to replicate the views of people who have studied comparative risks widely in the general populace? Is it to enable the accomplishment of those social goals that would be endorsed by utilitarian or Kantian ethical theories, respectively? Even if one remains agnostic about the answers that might be given to such questions, the above analysis suggests that specialists in risk communication would be well advised to study the relationship between utilitarian and Kantian ethical theories more carefully in the future. At a minimum, simply presuming that the least ethically complex cases of risk communication are prototypical is unjustified.

Grammatical Bias in Risk Communication

Given its usage in ordinary language, the word "risk" displays grammatical patterns that deviate significantly from the definitions in use by scientific analysts. Two issues in expert versus lay usage can take on ethical significance. One concerns whether or not all references to risk imply some sort of value judgment, or whether some standard of strong value neutrality is possible. The second concerns the way that many ordinary language discourse contexts make strong associations between risk and agency, implying an opening to further discourse on responsibility. In contrast, the usage specified in scientific risk assessment makes no association with intentional action and can be applied readily to circumstances in which human beings, organizations, or other intentional agents play no role in creating the conditions for risk. Each of these issues is discussed in turn.

Some theorists have argued that the scientific analysis of risk should be "value free," while acknowledging that any attempt to manage risks will inevitably involve a decision maker in value judgments (Rowe, 1977; Cross, 1998). Others have countered that any conceptualization of risk will involve at least the value judgment that possible outcomes are regarded as adverse. On this latter view, people do not discuss the risk of happiness and satisfaction unless they are speculating on the possibility that these situations might have some unnoticed downside potential (Schulze & Kneese, 1981; Rayner & Cantor, 1987; Hansson, 1996; Cranor, 1997). As noted above, Paul Slovic, a founding figure in the study of risk perception and risk communication, has characterized the divide between these perspectives as "the risk wars" (Slovic, 1999).

Only recently have empirical studies on the way non-specialists talk about risk begun to appear. Tulloch and Lupton (2003) conducted a number of interviews on the topic of risk and report their findings in the book *Risk and Everyday Life*. They report

> a dominant tendency to characterize risk as negative. The emotions of fear and dread were associated with interpretations of risk as danger and the unkown. Uncertainty, insecurity and loss of control over the future were associated with risk, as was the need to try and contain this loss of control through careful considerations of the results of risk-taking. But there was also evidence in many people's accounts of positive meanings associated with risk: adventure, the conditions of excitement, elation and enjoyment, the opportunity to engage in self-actualization and self-improvement. (p. 19)

Tulloch and Lupton undertook their studies in the context of evaluating Beck's thesis that a feeling of being at risk has become pervasive in modern society (1992). They believe that while aspects of their work support Beck's thesis, counter-themes associated with positive meanings reinforce the sense that meanings of risk are subject to significant variation depending on race, gender, and cultural location. Their research suggests that in ordinary discourse, risk is inherently value-laden. A quantitative study of numerous examples of discourse using the word "risk" does not support this generalization, however, and concludes: "From the frame semantics perspective, linguists found that at its core, as both a noun and a verb, 'risk' emphasized actions, agents or protagonists, and bad outcomes such as loss of a valuable asset" (Hamilton, Adolphs, & Nerlich, 2007, p. 178).

My own view is that value judgments are involved, but that in many cases they are definitive and hardly noticed. For example, in many areas of risk assessment relevant to public health the value judgment implied by taking human mortality and morbidity to be a bad thing is utterly uncontroversial. To point out that it is nonetheless a value judgment is not to imply that it should be debated. Furthermore, advocates of a so-called value-free risk science are clearly correct to insist on the relevance of science both to the identification of hazards and to the derivation of probabilities to estimate the likelihood that harm from hazards will actually materialize. The scientific analysis of these elements should be shielded from forces that would bias the analysis toward favored outcomes. To say that science should be shielded from biasing forces expresses a value, albeit an epistemic value rather than a sociopolitical value. Ben-Ari and Or-Chen (2009) draw a distinction between social values, which express preferences toward specific social outcomes, and epistemic values, which specify the norms for scientific inquiry. They argue that recognizing the difference between these two types of value will clarify a number of disputes in scientific risk assessment.

However, over the last several decades it has been more typical for risk science to advocate a "value-free" ideal. The "value-free science" viewpoint is often conjoined with an insistence on the scientific community's authority to specify the meaning of words. To insist that risk, properly understood, must *always* be understood as a function of hazard and exposure illustrates a grammatical bias with ethical significance, but it is important to understand why scientific risk analysts take such pains to talk this way. Within the context of scientific risk assessment, clear definitions of hazard, or the potential for harm that may be associated with a phenomenon or activity, and exposure, or the factors that contribute to the materialization of this harm, are essential. Once characterization of hazard and exposure are available, it is possible to interpret the risk associated with a possible event *e* according to the broadly specifiable formula

$$\text{Risk}_e = f(v_e, p_e)$$

where v_e is the harm or harmful outcome (often specified simply in terms of morbidity and mortality) and p_e is the probability that the event occurs, given scientifically investigable conditions of exposure. In this formula f can incorporate a number of complex mathematical concepts. In the simple case of quantifying the risk of losing a standard coin flip, v_e is the amount wagered, p_e is the probability of losing (i.e., 0.5) and f is multiplication, so the expected adverse value (i.e., risk) is one half of v.[2]

The simple coin flip illustrates the communicative challenge. If you ask someone what they think the risk of a coin flip is, they might well respond

2. The positive outcome (e.g., winning the flip) is one half of $+v$, making the total expected value of a standard coin flip zero.

by saying that it is what they stand to lose (i.e., v, the amount wagered). But this quantity does not reflect the fact that their *chance* of losing is actually 0.5. The quantitative risk associated with a standard coin flip is one half of the amount wagered, thus not equivalent to whatever one stands to lose. When complex activities embroil the calculation of probabilities in lengthy causal chains and uncertainties, this basic communication challenge is heightened. Scientists who have undertaken painstaking work to estimate hazard and exposure are, in one sense, understandably frustrated when the word "risk" is used in popular contexts in ways that connote the mere potential for harm, with little or no recognition of its likelihood.

In ordinary language the word "risk" is used in this way all the time. However, there is a more subtle point. In ordinary language discourse, the word "risk" can function as a verb. In this grammatical usage, it connotes action by an agent capable of intentionality: "John and Jane risk their personal fortune with an ill-considered business venture." "The Hundsucker Corporation risks the health of everyone in Springfield with its new chemical plant." In these instances, a person and an organization are spoken of as doing something. But in sentences with non-agential subjects, using the word "risk" as a verb produces grammatically ill-formed constructions that must be interpreted metaphorically, if they can be parsed at all: "The mountain risks its flora and fauna with an earthquake." "The tornado risked life and limb in Springfield." Individuals and groups are intentional agents, mountains and tornados are not. There are indeed borderline cases. Animals and other living things can be described as if they acted intentionally: "The mother bear risks her cubs by straying too far." "The begonias risk a late frost by blooming early." But the larger point is that the formula $Risk_e = f(v_e, p_e)$ does not express an action at all (Thompson, 1999, 2007).

The authors of the quantitative study cited above note similarity between the core properties of the word "risk" when used as either a noun or a verb, and remark on the "interesting" fact that the word is used as a noun much more frequently in scholarly databases than in databases including usage from non-scholarly contexts. This quantitative study supports a correlation between the use of the verb "risk" in discourse contexts in which agency is involved, as distinct from natural hazards or random events. This finding contradicts the view that risk is always adequately characterized by a formula such as $Risk_e = f(v_e, p_e)$. Expressed as the mere probability of a harmful outcome, risk is tied conceptually to causality, but not agency. I have suggested that there is a natural flow to risk discourse in which early messages that bring risk to the attention of the audience are interpreted according to an "act-classifying" sense of the word. On this interpretation, a key conversational function of the word "risk" is to move the topic under discussion from the category of the unexceptional and quotidian into the category of those topics, events, or activities that call for action by someone, often the person to whom the communication is directed. Once the topic has been accepted as one calling for action, it may become pertinent to consider the details of probability more carefully. In these contexts, an "event-predicting" sense of the word "risk" that is quite consistent with the $Risk_e = f(v_e, p_e)$ formula preferred by experts may take over. It is not as if ordinary people never think or talk about risk in the same way as experts. Rather, we should expect that any message that interrupts the normal flow of events to raise the subject to risk will be interpreted as a signal that some sort of action needs to be taken (Thompson, 1999; Thompson & Hannah, 2012).

Other work from risk communication studies provides indirect support for a view that I developed from the philosophical study of the way that discussions of risk organize our thoughts toward moral and prudential responsibilities. Risk communication scholars working to effect behavior change have struggled with problems that arise when audiences react to messages fearfully or take them to threaten their self-concept (Lapinski & Boster, 2001; Covello & Sandman, 2001). Witte has proposed a model that emphasizes attention to efficacy—that is, the ability of a recipient of a communication message to take action to remediate or otherwise address the risk (Witte & Allen, 2000; Cho & Witte, 2005). A grammatical analysis suggests participants expect a conversation on risks to involve some point of action that they or others should initiate. There would be no point to a risk communication focus on, for example, background rates of radiation exposure or the eventual death of the sun.

The implicit grammatical tie to agency explains why communicators err when they raise the topic of risk (even implicitly) only for the purpose of reassuring people that everything is just fine. The annals of GMOs provide examples, once again. Risk messaging on foods containing genetically engineered ingredients was calculated to assure consumers that these foods were "substantially equivalent" to other items on their grocery shelves. The message that many took from this messaging was quite the opposite (Katz, 2001). In sum, as with ethical theory, philosophical work in ethics (s.) on the structure and context of risk-oriented discourse and decision making provides the basis for both insight and new hypotheses. Scholars of language and communication might well contribute to an improved understanding of the risk communication processes by paying closer attention to the way that the word "risk" is used in ethically oriented conversational contexts.

Conclusion

Risk and risk communication get inevitably tangled up in ethical consider-ations because to call something by the name risk is to imply that people have reason to avoid it or at least to be mindful about it. One is thus imply-ing that they have a reason to expect that they (or someone) has something to lose in regard to the topic under discussion. According to one common way of speaking, this is a matter of prudence, rather than ethics. However, in ethics (s.) all topics in which matters of "should" and "ought" arise are of interest. And there are many cases in which the topic of risk does involve how one's action affects others and vice-versa, in any case, so even by the most exacting standards, conversations about risk are very likely to involve ethical considerations. On the one hand, it is striking how little attention is given to ethics (s.) in the literature of risk communication. For the most part, those scholars who study attitude formation and behavior change in communica-tion practices related to risk display virtually no interest in the ethical ques-tion of whether using communication techniques to influence attitudes and change is ethically justified. On the other hand, this fact may not be so strik-ing when one realizes that many of the risk communication efforts that have been mounted clearly are justified and could be shown to be justified using any of the strategies for critical thinking that ethics (s.) provides. Neverthe-less, I hope that I have shown that such happy conditions of easy justification are not universally the case.

Utilitarian and Kantian moral frameworks present somewhat different ways of thinking about risk and risk communication. This suggests that in some cases, we think that getting trade-offs right is the goal, while in others we think that the goal of analyzing and then communicating about risk is simply one of placing people in a position where they can make their own decisions. There are also cases where these two goals get tangled up and it becomes controversial (or perhaps simply confusing) as to what risk com-munication is trying to achieve. On top of this, there are tangles and confu-sions that arise in connection with the message that risk communication is to contain, regardless of its overarching purpose. Are risk messages simply supposed to tell us how the world works so that we can better estimate the probability of harm or loss? Or are they intended to grab our attention, shake us by the shoulders, and engage in a deliberative search for what we should do? Although forays into ethics (s.) probably do too little to help would-be risk communicators and scholars of risk communication answer these ques-tions, it may nonetheless be important to shake them by the shoulders a bit and provoke a bit more thoughtful and deliberative inquiry into the nature and function of communicating scientific messages about risk.

References

Appiah, K. A. (2008). *Experiments in ethics*. Cambridge, MA: Harvard University Press.

Beck, U. (1992). *Risk society: Towards a new modernity*. London: Sage.

Ben-Ari, A., & Or-Chen, K. (2009). Integrating competing conceptions of risk: A call for future direction of research. *Journal of Risk Research, 12*(6), 865–877.

Bentham, J. (1948). *An introduction to the principles of morals and legislation*. Introduction by Laurence J. Lafleur. New York, NY: Hafner. (Original work published 1823)

Bradbury, J. A. (1989). The policy implications of differing concepts of risk. *Science, Technology & Human Values, 14,* 380–399.

Cates Jr., W., Grimes, D. A., Ory, H. W., & Tyler Jr., C. W. 1977. Publicity and public health: The elimination of IUD-related abortion deaths. *Family Planning Perspectives, 9,* 138–140.

Covello, V., & Sandman, P. M. (2001). Risk communication: Evolution and revolution. In A. Wolbart (Ed.), *Solutions to an environment in peril* (pp. 164–178). Baltimore, MD: Johns Hopkins University Press.

Cho, H., & Witte, K. (2005). Managing fear in public health campaigns: A theory-based formative evaluation process. *Health Promotion Practice, 6*(4), 482–490.

Cranor, C. F. (1990). Some moral issues in risk assessment. *Ethics, 101,* 123–143.

Cranor, C. F. (1997). The normative mature of risk assessment: Features and possibilities. *Risk: Health, Safety and Environment, 8,* 123–136.

Cross, F. B. (1998). Facts and values in risk assessment. *Reliability Engineering & System Safety, 59,* 27–40.

Derrida, J. (2008). *The animal that therefore I am (following)*. M. L. Mallet (Ed.), D. Wills (Tr.). New York, NY: Fordham University Press.

Eriksen, M. P., & Cerak, R. L. (2008). The diffusion and impact of clean indoor air laws. *Annual Review of Public Health, 29,* 171–185.

Ezra, D. B. (1990). Smoker battery: An antidote to second-hand smoke. *Southern California Law Review, 63,* 1061–1122.

Fan, D. P., & Holway, W. B. (1994). Media coverage of cocaine and its impact on usage patterns. *International Journal of Public Opinion Research, 6,* 139–162.

Green, D. P., & Gerken, A. E. (1989). Self-interest and public opinion toward smoking restrictions and cigarette taxes. *Public Opinion Quarterly, 53,* 1–16.

Hacking, I. (2001). *The social construction of what?* Cambridge, MA: Harvard University Press.

Hamilton, C., Adolphs, S., & Nerlich, B. (2007). The meanings of "risk": A view from corpus linguistics. *Discourse & Society, 18*(2), 163–181.

Hansson, S. O. (1996). What is philosophy of risk? *Theoria, 62,* 169–186.

Hansson, S. O. (2007). Risk and ethics. In T. Lewens (Ed.), *Risk: Philosophical perspectives* (pp. 21–35). New York, NY: Routledge.

Hauser, D. E. (1974). Is freedom of choice worth 800 fatalities a year? *Canadian Medical Association Journal, 110,* 1418–1426.

Hirst, E. W. (1934). The categorical imperative and the Golden Rule. *Philosophy, 9,* 328–335.

Irwin, A. (1987). Technical expertise and risk conflict: An institutional study of the British compulsory seat belt debate. *Policy Sciences, 20,* 339–364.

Jones, J. (1993). *Bad blood: The Tuskegee syphilis experiment* (new and expanded ed.). New York, NY: Free Press.

Kant, I. (2002). *Groundwork for the metaphysics of morals.* A. Zweig (Tr.), T. E. Hill Jr. & A. Zweig (Eds.). New York, NY: Oxford University Press. (Original work published 1785)

Katz, S. B. (2001). Language and persuasion in biotechnology communication with the public: How to not say what you're not going to not say and not say it. *AgBioForum, 4,* 93–97.

Killingsworth, M. J., & Palmer, J. (1992). *Ecospeak: Rhetoric and environmental politics in America.* Carbondale: Southern Illinois University Press.

Lapinski, M. K., & Boster, F. J. (2001). Modeling the ego-defensive function of attitudes. *Communication Monographs, 68,* 314–324.

Lewis, H.W. 1990. *Technological risk.* New York, NY: W. W. Norton.

Love, N., & Ansaldo, U. (2010). The native speaker and the mother tongue. *Language Sciences, 32,* 589–593.

Machan, T. J. (1984). Pollution and political theory. In R. Regan (Ed.), *Earthbound: Introductory essays in environmental ethics* (pp. 74–106). New York, NY: Random House.

Macklin, R. (1992). Universality of the Nuremberg Code. In G. J. Annas & M. A. Grodin (Eds.), *The Nazi doctors and the Nuremberg Code* (pp. 240–257). New York, NY: Oxford University Press, pp. 240–257.

Priest, S. (2009). Risk communication for nanobiotechnology: To whom, about what, and why? *Journal of Law, Medicine & Ethics, 37,* 759–769.

Rawls, J. (1980). Kantian constructivism in moral theory. *Journal of Philosophy, 77,* 515–572.

Rayner, S., & Cantor, R. (1987). How fair is safe enough? The cultural approach to societal technology choice. *Risk Analysis, 7,* 3–9.

Rollin, B. E. (2006). "It's my own damn head": Ethics, freedom and helmet laws. In B. E. Rollin, C. M. Gray, K. Mommer, & C. Pineo (Eds.), *Harley-Davidson and philosophy: Full-throttle Aristotle* (pp. 133–144). New York, NY: MJF Books.

Rothman, D. J. (1982). Were Tuskegee and Willowbrook "studies in nature"? *The Hastings Center Report, 12*(2 [April]), 5–7.

Rowe, W. D. (1977). *An anatomy of risk.* New York, NY: Wiley.

Russell, M. (1987). Risk communication: Informing public opinion. *EPA Journal, 13,* 20–21.

Schulze, W. D., & Kneese, A. V. (1981). Risk in benefit-cost analysis. *Risk Analysis, 1,* 81–88.

Singer, P. (1975). *Animal liberation.* New York, NY: Avon Books.

Slovic, P. (1999). Trust, emotion, sex, politics, and science: Surveying the risk-assessment battlefield. *Risk Analysis, 19,* 689–701.

Starr, C. (1972). Cost-benefit studies in socio-technical systems. In Committee on Public Engineering Policy (Ed.), *Perspectives on benefit-risk decision making* (pp. 17–42). Washington, DC: National Academy of Engineering.

Starr, C., & Whipple, C. (1980). Risks of risk decisions. *Science, 208,* 1114–1119.

Syme, S. L., & Alcalay, R. (1982). Control of cigarette smoking from a social perspective. *Annual Review of Public Health, 3,* 179–199.

Thompson, P. B. (1999). The ethics of truth-telling and the problem of risk. *Science and Engineering Ethics, 5*(4), 489–510.

Thompson, P. B. (2002). Why food biotechnology needs an opt out. In B. Bailey & M. Lappé (Eds.), *Engineering the farm: Ethical and social aspects of agricultural biotechnology* (pp. 27–44). Washington, DC: Island Press.

Thompson, P. B. (2007). "Norton's sustainability: Some comments on risk and sustainability." *Journal of Agricultural and Environmental Ethics 20,* 375–386.

Thompson, P. B., & Dean, W. (1996). Competing conceptions of risk. *Risk: Health, Safety & Environment, 7,* 361–384.

Thompson, P. B., & Hannah, W. (2012). Novel and normal risk: Where does nanotechnology fit in? In D. Schmitz & E. Willett (Eds.), *Enviromental ethics: What really matters? What really works?* (pp. 609–622). New York, NY: Oxford University Press.

Tulloch, J., & Lupton, D. (2003). *Risk and everyday life.* London, UK: Sage.

Vaccari, A. (2008). Prudence and morality in Butler, Sidgwick and Parfit. *Etica & Politica, 10*(2), 72–108.

Vogt, D. U., & Parrish, M. 1999. *Food biotechnology in the United States: Science, regulation, and issues.* Congressional Research Service Report RL 30198. Washington, DC: US Government Printing Office.

Walsh, D. C., & Gordon, N. P. (1986). Legal approaches to smoking deterrence. *Annual Review of Public Health, 7,* 127–149.

Watson, G. S., Zador, P. L., & Wilks, A. (1981). Helmet use, helmet laws and motorcycle fatalities. *American Journal of Public Health, 71,* 297–300.

Witte, K., & Allen, M. (2000). A meta-analysis of fear appeals: Implications for effective public health campaigns. *Health Education & Behavior, 27*(5), 591–615.

Wynne, B. (1996). May the sheep safely graze? A reflexive view of the expert-lay knowledge divide. In S. Lash, B. Szerszynski, & B. Wynne (Eds), *Risk, environment & modernity: Towards a new ecology.* London: Sage.

Yanovitzky, I., & Stryker, J. (2000). Mass media, social norms, and health promotion efforts: A longitudinal study of media effects on youth binge drinking. *Communication Research, 28,* 208–239.

3

Communicating Climate Change and Other Evidence-Based Controversies: Challenges to Ethics in Practice

S U S A N N A   P R I E S T

Climate change is warming up the world. The consequences have been well documented: not just changed temperatures, but rising seas, eroding and flooding coasts, the spread of invasive species, the disappearance of other species, chaotic weather, melting icecaps, reduced snowpacks, shifting patterns in drought and rainfall conditions, rises in the numbers of wildfires, and the likelihood of crop and livestock maladaptation to new conditions. Yet communicating the reality of climate change has proven much more difficult than might have been predicted. Communicating this reality and the need for action about it to a broad range of publics is vitally important if adoption and mitigation strategies are to be supported and implemented. However, in practice, what we must do to accomplish this goal both raises ethical issues and challenges accepted ethical principles. This discussion attempts to elucidate the points of tension.

Political ideologies and gridlock among political factions, especially problematic in the United States although also present elsewhere, are but one factor underlying this dynamic (Pearson & Schuldt, 2015). Although research has documented extensive funding being channeled toward fueling and legitimizing climate change "denier" attitudes (Brulle, 2014), much more than politics is to blame. Climate change is a reality that is stunning to contemplate. How could our earth be changing into an environment that will not support us in the ways we take for granted, and with such apparent rapidity? And it has become common knowledge among science communication specialists that presenting "just the facts" is rarely enough to change opinions or attitudes on controversial topics. Factors such as trust, credibility, consistency with prior beliefs, patterns of opinion leadership, social group membership, and values are also at work, along with fear and denial. Sensitivity to these influences calls for more than a better understanding of how best to construct a persuasive message targeted at specific demographic groups—much more than a "science of science communication" (National Academy of Sciences, 2013) might imply, in other words. The idea of using the same manipulative strategies to "sell" science (Nelkin, 1995) that we often use to sell products is, prima facie, not entirely consistent with the idea of a pluralistic democracy in which people make up their own minds about things, especially when the government and the scientific community are the "sellers" and the ordinary citizen is the potential "buyer." It is also not very likely to work.

Ethical traditions are thus also in play—and in flux. This chapter outlines some of the particular ethical challenges that communicating the reality of climate change presents for both journalists and scientists, as well as for other science communicators, such as educators and concerned citizens (including those who do and those who do not self-identify as environmental "advocates"). This discussion builds on two concepts initially developed in earlier work. One of them is the tension between science communication as *serving strategic goals* and science communication as *serving democratic goals.* The second concept is related to what I call "critical science literacy" (Priest, 2013a), a type of literacy that both citizens and journalists need in order to make sense of messages about science in a social climate in which those messages may serve either strategic or democratic objectives (or, arguably, in some cases both—the lines between them are sometimes blurry in practice). Critical science literacy is also important for other science communicators (including scientists themselves) to keep in mind. Without critical science literacy, audiences are unlikely to be able to distinguish legitimate scientific uncertainty and disagreement from false controversies—including those presented in order to mislead. Democratic objectives related to science and science policy are unlikely to be fulfilled under conditions in which critical science literacy is low.

Strategic versus Democratic Communication

Both journalists (especially in the United States, historically) and scientists generally operate within systems of professional ethics that privilege reliance on "objective" facts over personal "opinions," particularly for science reporting (Nelkin, 1995). In short, neither group is traditionally supposed to "take sides" on public or policy controversies (although the difference between "objective fact" and "personal opinion" can sometimes be problematic, even when the topic in question is a scientific one).[1] Some science communication is strategic (serving the strategic interests of the communicator or those the communicator represents). Some is intended more democratically (serving the interests of democracy and its citizens by enabling informed decisions about science-related interests). In practice the two often overlap or are intertwined. This in itself is a challenge for all communicators.

A *strategic* science communication goal generally involves an active intent to accomplish something with respect to what people think (and do), whether it involves a health-related choice such as vaccination or tobacco smoking or promotes a corporate interest such as securing support for crop biotechnology or an advocacy interest such as preserving old-growth forest. Corporate and advocacy interests regularly fail to align with one another, for example when development goals (e.g., building an oil pipeline) conflict with conservation goals (e.g., preserving animal migration routes). However, spokespersons for these interests are not the only strategic communicators. Attempts to improve generic public approval ratings for science (ultimately tied, presumably, to levels of funding and other forms of public support) are predominantly strategic. This often takes the form of public relations and public information activities (despite the latter term having more of a "neutral" or "objective" tenor) on behalf of universities, research agencies, and other research institutions. Corporations and advocacy groups also use "neutral" scientific research in the service of their particular strategic interests. Any of these organizations can use advertising campaigns to support their strategic goals as well, and many do use this paid-for form of strategic communication, but providing information to journalists is also a key strategy for achieving strategic communication goals. When journalists receive research-related press releases, it is part of their job to ask whose interests they serve.

Strategic science communication goals are not always unethical, but they are always worth scrutinizing. Further, sometimes strategic goals and democratic principles coincide rather than collide. Direct-to-consumer (or DTC) pharmaceutical advertising comes to mind. While, on the one hand, it is hard for individuals to resist self-diagnosing on the basis of such ads, it can also be empowering for individuals to become aware of new vaccination options and other new treatments. We no longer live in a society in which a doctor is likely to be proactively taking care of us as individuals—or would, for example, recognize us in the grocery store. Perhaps it is not wise to cede all

---

1. One can, of course, use "objective" information strategically, and the expression of "opinion" is an important element of democracy.

health management power to the individual, but de facto that is often the way it is. For the most part, we need to proactively go to a doctor when troubling symptoms arise or we learn that a new option is available. Do DTC drug ads help good decision making by citizens more than they hurt it? We can readily imagine people becoming overmedicated as a result of this advertising, to no good end. Alternatively, we can also imagine them suffering in silence because they had no idea their symptoms were problematic or could be treated. There is not an easy answer as to which is the greater risk.

Democratic science communication goals, on the other hand, are those involving a central intent to improve how well democracy works for science-related issues and how well citizens in a democracy are able to make good decisions that are in their own self-interests and in the interests of society as a whole. Information about science is provided so that people can be informed, in other words, as they make up their own minds. The "mental model" this conjures up is of a base of undisputed truthful information available as a sort of data bank or giant library to society at large, where journalists and scientists are the gatekeepers while individuals then take up the neutral information they are provided and argue about its interpretation and application (at times strategically, to be sure). One problem with this vision is that many science-related issues are value-based, or based on nonscientific belief systems, such as one's position on stem-cell research, abortion, or human cloning. Nevertheless, one hopes that people will make up their minds on such issues while possessed of the basic scientific facts, rather than scientifically false information. Those scientific facts can, in theory, be offered in a neutral way.

The other big problem here is that all scientific "truth" is tentative: It is subject to revision, different scientists might interpret the same data differently, and uncertainty cannot be fully eliminated. There is no neutral data bank. There are, however, some scientific facts (and ideas) that are characterized by broad scientific consensus (such as the existence of climate change) and other scientific "facts" that are characterized by competing scientific interpretations (here, theoretical physics comes to mind). One might think of these latter as candidate future facts, but at present the uncertainties outweigh the certainties, a balance that is constantly changing. Other scientific assertions, such as the idea that vaccination causes autism, have been discredited. However, the concepts of relative balance between certainty and uncertainty and of consensus views within science are of course complex. How are ordinary citizens to weigh them? While scientific truth is subject to revision, I would not argue that it is entirely subject to popular referendum either.

Having introduced those caveats, some science communication efforts are certainly characterized by a heart-felt and completely sincere desire to empower citizens to make their own decisions. Given that climate change is practically certain to result in rising sea levels, a land-use planner who provides participants in a town meeting with that information probably wants them to do something about it but may have no particular stake in what sort of response is ultimately chosen. Given that exposure to loud sounds damages hearing, people exposed to them are advised to protect their ears—and those giving this advice are not necessarily in the business of selling a particular brand of ear plugs. However, relatively pure examples are not always easy to find. Given that some household chemicals are toxic, warning labels and poison control centers hope to persuade citizens to use appropriate caution and, if necessary, to get appropriate treatment quickly. In this case the message providers may have no personal stake in the outcome, but their communication can clearly still be seen as strategic (even if humanitarian): They want people to take specific, defined action.

Almost all messages do have some strategic element. The clearest distinguishing features are whether the communicator intends to put the power to make decisions in the hands of the message recipient and whether the information is known to be either false or incomplete in important, directly relevant, ways. While democratic and strategic messages are conceptually distinct, in practice this is a continuum.

Like the DTC drug ads mentioned above, many other science communication activities clearly support both types of goals. For example, educational anti-tobacco messages generally serve the strategic (persuasive) goals of both health agencies and certain advocacy groups. Further, if a university (for example) has researchers that have found improved methods to help people quit smoking, they will want journalists to cover it. This can generate a message (such as a press release and related news article) that both discourages tobacco use and promotes the university's research capabilities—both of these being strategic goals, although it may also be sincerely hoped that the information will empower intelligent choices. As a result of persuasive strategic messages, people might become both more motivated and more empowered to make positive personal decisions (in this case, by quitting smoking). Everyone should be happy about this—except tobacco growers and cigarette companies, for example, who will probably be busy devising other strategic messages involving things like the so-called freedom to smoke. Similarly, some segments of the energy industry (such as coal producers) argue on "free market" grounds that their activities should not be further regulated in response to climate change.

This kind of conflict between strategic and democratic goals constitutes an important underlying tension, especially to the extent these goals are not always entirely transparent (to message recipients, for example). Strategic messages are certainly not limited to those that are disseminated by purely selfish interests at the expense of others. Everyone who is trying to promote anything through communication is using the strategic form, and in a pluralistic democracy that means a good share of us.[2] Good democracy can be fed, in principle, by struggles among multiple strategic messages representing different points of view in what is sometimes called the "marketplace of ideas" model. In other words, strategic communication sometimes fits into democratic practice. Messages with purely democratic science communication goals might be rare in practice, although it is often the ideal form of communication practice that is implicitly assumed (even if rarely achieved). Most communicators are trying to accomplish something other than neutral education. At the same time, both journalists and scientists are often idealistic about communication goals, committed to the ethical and other professional principles of their respective fields as well as to the broader society. And in an ideal world, both might operate completely independent of particular special or personal interests and provide only disinterested information. This is not so easy as it might sound, of course. Almost no information is completely neutral.

Even so, people in either profession might tend to conceptualize their work (with respect to communication with broader publics) as a matter of identifying truthful information and presenting it in an easily digestible form, such that rational people will make their own decisions, as democratic theory assumes that they will. In practice, however, purely democratic communication may not be as ubiquitous as strategic communication, even in the work of journalists and scientists. Journalists are regularly accused (sometimes with justification) of doing their work, self-servingly, in a way that will appeal to their audiences, through tactics such as over-sensationalization. Scientists are sometimes accused (occasionally with justification as well) of picking evidence to fit their theoretical preconceptions. In some cases, the commitment to an idealized vision of the purpose of public communication as serving the public can leave both communicators and message recipients somewhat oblivious to underlying, perhaps even unintended, strategic dimensions—and effects.

A further complication that is often set aside and not fully unpacked is that two equally "rational" people, faced with the same good information,

---

2. Of course communication, including science communication, can have other purposes not considered here. For example, it can also have entertainment value (and associated economic value). Entertainment can itself be strategic/persuasive and/or democratic/educational—or, hypothetically, it can simply be entertaining.

can and do make different choices. But this is very often the case, especially in the kind of globalized, pluralistic society we almost all live in today. Values, not just "proper" information, underlie most personal and policy decisions. And a lot of information is simply not neutral, however packaged. Still, it is useful for the sake of discussion to imagine science communication on a spectrum from information that serves primarily the interests of the message recipient for purposes of democratic participation (empowering understanding and decision making, independent of judgments about what decision is best), an idealized type, and information that serves a narrower strategic goal and seeks to produce a particular outcome, particularly one that is seen as being in the strategic interests of the communicator or someone the communicator represents—rather than the interests of the message recipient.

Science Communication and Public Controversy

An important element of democracy is, of course, the freedom of citizens to make a variety of choices about their lives. In the United States, one of the core freedoms to choose concerns religious choice (protected, alongside freedom of speech, by the First Amendment to our Constitution). If, as a US citizen, I choose not to believe in evolution but instead to believe that God created the Earth, its organisms, and its environment in more or less their present form at a single historical point in time, I am exercising a constitutionally guaranteed right. Educating me to understand *what the idea of evolution is* (and what scientific evidence is available that supports it) may be democratic to the extent that this empowers me to comprehend popular debate on this point and to make an "informed decision" about whether I want to accept it as my own belief. Educating me for the purpose of persuading me to *change my beliefs* about evolution is more clearly strategic, rather than purely democratic, especially given that my right to believe what I want in the realm of religion is specifically constitutionally protected. The distinction is apparent in opinion poll results that differ depending on whether a question is about *what evolution is* versus when the question is about *whether I believe in it* (see, e.g., Rughinis, 2011). Yet many members of the scientific community (and communicators who work on their behalf) persist in the belief that providing scientific evidence will necessarily change opinions and beliefs. This is not always the case.

Similarly, it certainly seems democratic and empowering to educate people about the wisdom of vaccinating themselves and their children against preventable and potentially fatal diseases—and it is certainly not easy to argue otherwise. But this kind of communication is generally undertaken with a strategic goal; the communicator generally wants these individuals to change their behavior and embrace vaccination. This is desirable "for their own good," of course—as well as their children's good, *our* good (and our own children's good), society's good, and the global good, as well as the good of the healthcare industry and even the good of the companies that manufacture and sell vaccines. This very last dimension can be seized on by anti-vaccination advocates as evidence that attempts to persuade them to vaccinate are not purely disinterested, however, further complicating the opinion dynamics. Because the health and even the lives of children are at stake, it is not too difficult for many to accept that attempts to promote vaccination are morally justified, just as persuading people to stop smoking also seems morally justified. In such cases, a particularly good argument can be made that strategic communication may not always be unethical.

This presumption of moral good is based partly on the implicit assumption that long life and good health are universal values. Of course, good parents want their children to be healthy! However, there are individuals and groups who do not see things in quite this simple a way. The situation becomes much more problematic when rejection of vaccination (for example) appears to have a religious foundation, because of the philosophy of religious freedom embodied in the First Amendment—as opposed to being a matter of simply misunderstanding the relevant science or of having been exposed to "bad" science or being unable to trust the communicator. And yet all these things can also be mixed together. It is worth highlighting here as well that rejecting vaccination and rejecting evolution are not the same kind of case; on the basis of the best science we have, rejecting vaccination has direct and potentially severe consequences that rejecting evolution does not. Such considerations affect our intuitive moral judgments; we often apply a utilitarian (consequence-based) metric for measuring morality, whether we do so consciously or unconsciously.

Situations of public controversy over science can be subtle and delicately nuanced. The world is not divided neatly into "pro-science" and "anti-science" camps. Considering some of the nuances here has at times brought me to the point of questioning whether strategic communication (even if it has the goal of promoting something scientific) is, in practice, ever entirely compatible with the ideal of democratic communication—that is, communication truly designed to empower audience members and foster independent decision making, rather than simply persuading audiences to adopt a particular idea or position (see Priest, 2013b). I have met quite a number of scientists who see science communication as a strategy for getting others "on their side," and yet at the same time they seem to feel that the information they have to offer is (or should be) entirely neutral. This is a deep tension in the field of science communication, but one that is not commonly recognized or discussed; neither scholars nor practitioners commonly question whether it is possible to do both—nor how often one sometimes masquerades as the other, i.e., we may think of ourselves as empowering independent thinking when actually we are attempting to persuade others to our specific view of things. Yet, in the end, strategic goals cannot be stamped out like wildfire; they are everywhere—and, as noted, sometimes they seem ethically justifiable.

But it can be a dangerous mix, even so. Journalists crafting "objective" news stories run the risk of supporting strategic communication efforts even when they think of themselves as committed to supporting only democratic communication goals. Examples are easy to come by; it is all too easy for any of us to imagine a health journalist (for instance) reporting the effectiveness of a new treatment (say, a new "miracle" drug) without giving enough thought to whose strategic interests that information serves. Of course, there are common journalistic practices such as seeking out alternative expert opinions that may help guard against the bad consequences of the over-promotion of fringe science or other unsupported claims. Nevertheless, the single-study press release remains a staple of science communication efforts, and the easy way out of seeking alternative expert opinions—as in the case of climate change, where skeptical voices have too often been presented as "speaking for" a legitimate segment of the scientific community—can have bad consequences for society.

Scientists, most of whom are also committed (in principle) to "objective" truth, run a gamut of related risks: They may be privileging, albeit unconsciously, the values of their funding sources, their personal prejudices, their unique professional training, or theories about the way the universe works to which they are otherwise committed—but which cannot be claimed to have been finally proven. Indeed, in principle, all science is subject to refutation by future work, and this very often happens. True, at any given point in time, science can provide the best information available, but it can rarely be described as presenting an absolute representation of "the truth." Thus, the goal of providing only "neutral" scientific information is almost unachievable, in practice.

This notion of the difference between the "best available truth" and some sort of ultimate truth has been a sticky issue for climate change in particular. Our ideal principles of democracy—involving discussion and decision making among citizens with the freedom to choose and the power to act—assumes that those citizens share an underlying worldview and possess skills that allow them to interpret scientific (as well as other) information in an informed way. In the particular case of climate change, the probabilistic nature of the relevant science and the traditions of journalistic "objectivity" both create openings for strategic messages designed to sow doubt. That is particularly worrisome to the extent some audiences can easily be misled, and yet arguing that citizens in a modern democracy should not be empowered to make up their own minds about science-related issues sounds distinctly anti-democratic. Perhaps we should distinguish between an issue like climate change, which may constitute a science communication emergency because our window of opportunity for turning things around appears (based on the "best available" scientific evidence) to be closing, and other areas where science education can more safely wait for more extended discussion. This sense of immediacy also colors our moral judgments. Climate change is thus more like vaccination and less like evolution; immediate and partially irreversible harm is taking place.

Critical Science Literacy

The tentative conclusion of my earlier work on the tension between strategic and democratic science communication goals was that strategic science communication is not *necessarily* incompatible with democratic science communication under some circumstances, one of which is that we can assume audiences possess at least some level of critical science literacy. In short, for strategic and democratic goals to comfortably coexist, we have to assume that audiences are not easily fooled. To satisfy this condition, people need to understand a little bit about the sociology of science (how science works as a set of social practices) and the philosophy of science (how it is done, e.g., its methods and ultimate limitations). This is a very different concept from the old view of science literacy as acquaintance with a preset inventory of scientific facts, generally measured with multiple choice or true/false tests (the very tests that have often muddied the waters further by including such items as belief in evolution, as opposed to understanding of what evolution is). Modified goals for science education and science communication are implied here. After the educational philosopher John Dewey, I do believe that non-experts can form reasoned opinions about complex matters given necessary information (including information from technical experts), and that the success of democracy rests on this. But what information is necessary, and how are those non-expert citizens to sort the wheat from the chaff, so to speak, in a complex information environment in which messages serving strategic and democratic goals are routinely comingled?

Critical science literacy is something that journalists and other profession-als writing and otherwise communicating about science also need. Indeed, as the current economic restructuring of the news media industry continues to reduce the numbers of employed professional journalists (and may cut the numbers who specialize in science, health, environment, and other com-plex stories yet more steeply), critical science literacy among the audiences for messages about science becomes even more important. These audiences can less readily rely on news professionals as gatekeepers who will make reli-able distinctions between science and speculation. More information about science-related issues may be available than ever before, due largely to the development and proliferation of internet-based communication, but the confusion between democratic and strategic communication intentions is only being compounded. The current confusion over climate change serves as a potent example.

Critical science literacy involves many kinds of awareness and is not the type of knowledge that it is typically taught in science courses — or anywhere else specific, for that matter, including the role of values in science policy, the political and economic dimensions of science, scientific uncertainty and its inevitability, "human factors" such as error and bias (and even occasional fraud), the nature of scientific expertise and scientific consensus, the roles of theories and hypotheses, methodological diversity, and so on. And this is hardly a complete list, only an initial, suggestive one. Understanding science goes a long way beyond understanding scientific facts, in other words. If we are to debate issues related to science and science policy as a society, critical science literacy needs to be broadly shared. The more advanced and complex scientific knowledge becomes, the more apparent the need to enhance critical science literacy becomes. It may be that because of the need to act with relative speed, strategic communication may be more important for advancing public understanding of climate change than is purely democratic communication. (Arguably, the same might be said for certain health communication chal-lenges, such as quitting smoking or vaccinating children.) However, now is the time to start working on raising levels of critical science literacy for future generations.

Understanding and Accepting Climate Change

Climate change is a communication and policy challenge for the entire world (as well as a challenge to the concept of democratic science communication) in part because many issues related to the climate science involved require fairly strong critical science literacy skills to sort out. How is it that there can be a high level of scientific uncertainty *on details* (e.g., where there will be droughts and where there will be floods) and yet a reasonable level of scientific certainty *about the big picture* that anthropogenic climate change already actually exists and very much actually matters? Should observation and modeling be accepted as valid "methods," on a par with (say) experiments? What messages that might be distorting climate change communication spring from ideological, political, or economic (monetary) interests, rather than a sincere desire to help people reach the truth? It is hard to doubt the difficulty of answering these questions, on which—yes—even some "experts" might disagree, despite a strong international scientific consensus that climate change exists and that it is largely caused by human activity. Understanding the chemistry of the atmosphere and the nature of the carbon cycle (in other words, following the old deficit model approach to science education and communication that emphasizes factual knowledge) probably does matter, but has little potential to dispel some people's doubts, which are unlikely to be rooted in science knowledge deficits.

While a lot of research attention has focused on political ideology as a motivator for those who discount the scientific evidence on climate change, a number of other understudied but significant psychological barriers to accepting the reality and implications of climate change also exist. The idea that the earth is changing in a fundamental way, a way that could make it (or parts of it) a less hospitable home for human beings, is scary and therefore a difficult thing to accept. We know from health communication research that messages with a strong fear component can cause people to shut down, especially when they have a sense that there is nothing they can do (called "low efficacy" in the literature). The overall ineffectiveness of "fear appeals" has been confirmed for climate change (O'Neill & Nicholson-Cole, 2009). We also know from basic research in psychology that the cognitive dissonance that arises when someone is confronted with statements that are incompatible with strongly held beliefs motivates people to reject those statements.

In short, the reality of climate change is very difficult to reconcile with many people's basic understanding of the cosmos, their place in it, and their assumptions about the stability of physical reality, whether those assumptions are driven by a religion, a perspective on the environment, a taken-for-granted perception of reality, science, or simply "common sense." Arguably, more social science effort should be directed toward understanding these barriers, although cause-and-effect relationships between preexisting beliefs and responses to new information are not very easy to pin down—another area of research characterized by persistent scientific uncertainty.

Further, there is an unambiguous need for a reasonably speedy policy response at local, national, and international levels. This comes at a time when the political system in the United States seems completely gridlocked, economic interests are certain to oppose solutions that seem expensive, local governments often have reduced budgets, and international discussions of climate have become bogged down in questions of who wins, who loses, who is to blame, and who pays. Only very recently did the Obama administration in the United States start publicly reaching out on climate issues to the international community. The slow pace of this and other policy action is part of what makes climate an escalating science communication emergency.

Ethical Challenges for Climate Change Communication

Does this perfect storm of contextual factors change the ethics of science communication as applied to climate change and related issues (e.g., energy policy)? This discussion is only intended to pose, and cannot hope to completely answer, this question. What is clear is that the communication surrounding climate change presents challenges to a range of normative assumptions and professional principles. These challenges provide food for thought as science communicators move forward on both strategic and democratic fronts, and many of them help illustrate the role of critical science literacy. People's intuitions about what constitutes ethical communication in this area are likely to be shaped in part by perceptions of the urgency and severity of the emerging climate crisis. For those of us who believe this is an emergency and yet also believe in thoughtful public deliberation, it seems the current political polarization does not provide any easy path forward. And yet messages designed to persuade—perhaps even to manipulate—may not merely cross established ethical lines in some instances; they may also backfire, especially with audiences who have already made up their minds that climate change is some kind of environmentalist hoax. Both boldness and prudence are in order. This section attempts to further articulate the nature of these challenges.

FREE SPEECH

Democratic societies are generally built on the assumption that free and open debate ("free speech") will contribute to the distillation of truth and the development of wise policy so long as various points of view are freely put forward (e.g., by journalists). This is the "marketplace of ideas" concept. The protection of free speech is a basic legal principle in the United States and many other countries. For climate change, recent evidence confirms that truly substantial funding has been put behind climate denier voices in the form of conservative philanthropic foundation funding of "counter-movement" organizations, with the ultimate sources of this funding often hidden (Brulle, 2014). The communication marketplace for climate-related ideas can thus be understood as a distorted one in which voices that ultimately serve unacknowledged commercial or ideological ends can be both both subsidized and disguised. This is a direct challenge to the assumption that our system can be relied on to foster debate that is free, open, and transparent. Awareness of the potential for manipulation of science-related messages based on economic interests and ideological positions may help to limit our vulnerability to propaganda from denier sources, however. Being aware of such possibilities and therefore skeptical about one-sided messages is an aspect of critical science literacy.

BALANCED JOURNALISM

Professional practice in US journalism has long defined "objectivity" in terms of balance. The tradition may have been borrowed from political journalism, where many issues have a clearly defined left and right point of view, and it may have economic roots as well (since news that does not take a position can be marketed to the broadest possible audience). This same set of practices also characterizes science journalism. A good journalist is one who seeks alternative voices, and for science this means finding "expert" sources who are uncommitted to the latest findings and can help put them in context—perhaps even disagreeing with an interpretation—instead of relying on the interpretation of the researcher or even that of a press release writer at their institution. For climate change, however, this has sometimes meant casting deniers in the role of "opposition" voices, even "experts," creating confusion among audience members (Boykoff & Boykoff, 2004). Fortunately, there is evidence that this particular challenge has caused leading environmental journalists to rethink the nature of "objectivity" (Hiles & Hinnant, 2014).

PUBLIC ENGAGEMENT PRACTICE

The community of scholars concerned with science communication has largely abandoned the deficit model that defined the role of the science communicator as shaping attitudes by facilitating one-way transfer of scientific information from scientists to "non-expert" publics. Instead, we have become champions of "public engagement" alternatives in which scientists are encouraged to listen to their audiences, engaging in two-way dialogue, while citizens are encouraged to actively participate in discussions about—and sometimes the actual conduct of—science. This is certainly a positive and progressive trend. (See also Perrault 2013 for a recent discussion of this trend that urges us to move yet further away from the deficit model and toward the creation of a new, more democratic, relationship between science and society.) Unfortunately, the US town hall meetings about health care reform in the first term of the Obama presidency showed that broad societal discussion does not always produce progressive outcomes, at least in the short term. Even if it will do so in the longer term, our time to act on the slow-moving disaster of climate change is short and we cannot wait indefinitely for democratic discussion to take its course.

Of course, I do not disagree that it is vital public discussion should continue. And, fortunately, most Americans already believe climate change is happening. In addition to continuing to support ongoing public discussion of both climate science and energy policy alternatives, however, perhaps we can do more to consciously reinforce those beliefs, rather than fostering the misperception that the public is as highly polarized as the "dueling scientists" that so often appear in traditional journalism. By simply speaking out and making ourselves visible as climate "believers," we can actually weaken the denier minority and potentially strengthen the believer majority. Like scientists themselves, perhaps we should become more action oriented in our approach. This doesn't have to imply concerted activism, but can encompass simple acts of speaking out about the truth. The climate of public opinion is influenced by voices that are active and audible, not silent ones.

TRADITIONAL SCIENTIFIC ETHOS

Sociologist Robert Merton was one of the first well-known scholars to point out that science seems to have its own implicit professional code of ethics. One of the ideas many scientists take as an ethical principle (although not necessarily one of the ones Merton documented) is that scientists should remain, like journalists, "objective" and not take sides on policy controversies. Indeed, historically, scientists such as the late astronomer Carl Sagan who appear to have become "too" public have risked losing the respect of other scientists. I would agree that, under normal circumstances, some scientists might reasonably choose to avoid (as scientists) championing specific policy solutions, even on something like climate, lest they be mistaken for self-appointed (or media-appointed) authorities over what society should do. But scientists are also citizens, who may also reasonably choose to speak out, as citizens, on policy issues. To avoid involvement altogether can itself be seen as irresponsible. Scientists have the same rights and duties as other citizens, including the right and the duty to give thought to public issues and, on appropriate occasions, to take positions on them. For a climate scientist not to support appropriate action on climate might be likened to a medical doctor's not supporting vaccination programs, routine cancer screening, or prenatal nutrition initiatives. Nevertheless, this is a challenge to the traditional ethical thinking of many scientists who may feel it is important that they not "take sides" and in the process risk losing being seen as neutral and disinterested arbiters of truth.

Another point of conflict with the traditional ethos of science is the issue of transparency. It is understandable (and often a sore point in scientist-journalist relationships) that most good scientists are meticulous about specifying methodological limitations, levels of uncertainty, error bars, and other qualifications surrounding their conclusions. But widespread critical science literacy with respect to method—and particularly with respect to the concept of probability—cannot always be assumed. In areas like climate change where there is broad scientific consensus as well as high levels of apparent public controversy, high levels of media attention, and active popular interest, it might be that somewhat less transparency is actually called for. Scientists should not keep their data hidden, of course, nor should they distort it, but they should remember that when they talk about its inevitable shortcomings and the ever-present need for further research, they have audiences other than potential funding bodies and their own fellow scientists. Not every member of those audiences will understand that error doesn't always mean mistakes, or that the presence of uncertainty does not mean we have no basis for drawing conclusions. That science provides the best available evidence *despite* the presence of potential error and tangible uncertainly is a key component of critical science literacy.

THE ETHICS OF EXPERT INFLUENCE

It follows, then, that neither scientists nor science communicators nor science communication scholars have an obligation to keep silent on issues like climate change; rather, they may well have an obligation to speak out. Propaganda has many definitions, but one of the key characteristics of propaganda is that it attempts to change people's opinions in an "unfair" way—either through presenting one-sided material (which, ironically, rarely works well), by covering something up or misleadingly redefining it as another type of message (e.g., "disinformation"), or by somehow coercing audience members to adopt a particular position. To use one's social authority as a scientist or scholar in an overly heavy-handed attempt to win over others (especially on a topic to which that authority is actually not relevant) might be a form of coercion, but to act as an opinion leader on a topic relevant to one's expertise, or as a scientist-citizen even on a less immediately relevant topic, is not (in my view) out of bounds. If critical science literacy on the question of expertise were more widespread, there might be less of an issue here. However, as things stand, the responsibility should probably lie with the expert to clarify which positions are held as an expert and which as a regular citizen.

Intuitively, it seems that responsible communicators should strive to respect others' beliefs (even when contradicting or challenging them). This seems a basic tenet of respectful civil discourse. Climate change denial messages, on the other hand, have the potential to help directly to block and delay relevant policy action by exaggerating the uncertainty associated with still-emerging science and by keeping elected politicians aware of denier opposition, both of which provide handy rationales for political failures to act that may also please powerful economic interests. A lack of action now will have important and tangible bad consequences for future generations. Meanwhile, climate change believers who feel their side is losing the battle for public opinion may become less motivated to act or speak out themselves, a phenomenon public opinion researchers refer to as a "spiral of silence." Respecting alternative beliefs and interpretations need not entail that it is never ethical to speak out or to contradict.

So it is important that climate believers, including scientist and scholar believers, speak up. It is not inherently either disrespectful or manipulative to have an expert opinion and to express it in a way appropriate to the social context while hoping to win converts to the side of one's own truth-as-perceived. This is also the stuff of civil discourse. Not all persuasion is propaganda, in other words.

Conclusion

Some of the discussion above may recall for some readers, as it did for me, the well-known early twentieth-century disagreement between Walter Lippmann and John Dewey about the capacity for ordinary citizens to comprehend complex issues in a meaningful way. I remain on Dewey's side: I believe that given the right tools (and for science, I have grouped many of these necessary tools together under the "critical science literacy" umbrella), opportunities, and time, people will sort out the truth and democracy can be allowed to prevail. Unfortunately, climate change policy (like decisions in the foreign policy arena that had so strongly influenced Lippmann) is an area where we may not have enough time for these processes to play out without some degree of intervention by the community of experts—it is indeed a science communication emergency. I do not think this means that we have to resort to manipulation or half-truths (that is, to propaganda—a too-extreme version of "strategic communication"). It may mean, however, that scientists, scholars, and journalists concerned with climate issues have an obligation to be more proactive in helping to consolidate majority opinion, both within and outside of the scientific community, and making it more visible—and vocal.

This chapter would not be complete without a record here of my confession that every time I hear someone (say, someone in the grocery store line) mention odd weather, especially bad odd weather, I'm inclined to respond, "Well, that's climate change!" Of course, this isn't exactly true, and many public statements from the scientific community have stressed the difficulty of drawing cause-and-effect conclusions about the relationship between general climate patterns and specific weather, but then again it isn't exactly false either. My own goal is simple: to make people more aware of changed circumstances in their own immediate environments, including not only the weather itself but also the fact that some of their neighbors (me, for example) might be thinking of climate in observing the weather. This goal is both strategic and democratic.

References

Boykoff, M. T., & Boykoff, J. F. (2004). Balance as bias: Global warming and the US prestige press. *Global Environmental Change, 14,* 125–136.

Brulle, R. J. (2014). Institutionalizing delay: Foundation funding and the creation of U.S. climate change counter-movement organizations. *Climatic Change, 122*(4), 681–694.

Hiles, S. S., & Hinnant, A. (2014). Climate change in the newsroom: Journalists' evolving standards of objectivity when covering global warming. *Science Communication, 36*(4), 428–453.

National Academy of Sciences. (2013). The science of science communication. *Proceedings of the National Academy of Sciences of the United States of America, 100*(Suppl. 3), pp. 13696, 14031–14110.

Nelkin, D. (1995). *Selling science: How the press covers science and technology* (Rev. ed.). New York, NY: W. F. Freeman.

O'Neill, S., & Nicholson-Cole, S. (2009). "Fear won't do it": Promoting positive engagement with climate change through visual and iconic representations. *Science Communication, 30*(3), 355–379.

Pearson, A., & Schuldt, J. (2015). Bridging climate communication divides: Beyond the partisan gap. *Science Communication, 37*(6), 805–812.

Perrault, S. T. (2013). *Communicating popular science: From deficit to democracy.* New York, NY: Palgrave Macmillan.

Priest, S. (2013a). Critical science literacy: What citizens and journalists need to know to make sense of science. *Bulletin of Science, Technology & Society, 33*(5–6), 138–145.

Priest, S. (2013b). Can strategic and democratic goals coexist in communicating science? Nanotechnology as a case study in the ethics of science communication and the need for "critical" science literacy. In J. Goodwin, M. Dahstrom, and S. Priest (Eds.), *Ethical issues in science communication: A theory-based approach: Proceedings of the Third Summer Symposium on Science Communication, Iowa State University, May 30–June 1, 2013* (pp. 229–243).

Rughinis, C. (2011). A lucky answer to a fair question: Conceptual, methodological, and moral implications of including items on human evolution in scientific literacy surveys. *Science Communication, 33*(4), 501–532.

4

Framing Science for Democratic Engagement

LEAH SPRAIN

The very possibility that "framing" could be part of ethical science communication—not to mention warrant a chapter in a book on science communication ethics—is contested. In 2007, Matthew Nisbet and Chris Mooney published an article in *Science* arguing that scientists must learn to actively frame information to make it relevant to different audiences. Frames organize central ideas, defining a controversy so as to resonate with core values and assumptions. Frames pare down complex issues by giving some aspects greater emphasis. They allow citizens to rapidly identify why an issue matters, who might be responsible, and what should be done.

Discussing contentious issues such as climate change, evolution, and stem cell research, Nisbet and Mooney argued that scientists should use frames likely to engage broad public support because they resonate or at least complement existing worldviews. For example, climate change should be framed as a problem of religious morality or stem-cell research should be framed using "social progress" and "economic competitiveness" to forefront the hope that the research offers the public.

This article was met with significant skepticism, frustration, and lots and lots of commentary in the blogosphere and academy. Nisbet (2007) himself titled a subsequent blog entry, "Framing Science sparks a seismic blog debate." The firestorm over framing wrestled with fundamental questions about the role of scientists and science communicators within the public sphere, how people process information, public attitudes about science and scientists, how language works, and the ethics of science communication. This chapter will not rehash this entire controversy. But I use the social drama over the Nisbet and Mooney article to establish three related premises central to my argument about framing science communication.

First, the debates over the Nisbet and Mooney (2007) article demonstrate the need for an ethics of framing science communication. Framing was quickly associated with "spin." Scientists questioned whether framing is fundamentally lying—withholding relevant information, simplifying until science is inaccurate, and abandoning key aspects of science simply because it might not resonate with audiences. Questions were raised about whether casting science in nonscientific ways (e.g., religion) defeats the whole purpose of science—whether some frames meant that people are not, in fact, talking about the science. These challenges are important for science communicators, and they call for an ethics of framing science communication. Nisbet (2009) himself offered four ethical principles for framing science communication in a subsequent article. Whereas these principles provide a useful starting place, they are insufficient for supporting democratic engagement on public controversies over science, technology, and environmental issues, as I'll argue below.

Second, framing is inevitable. All communication is framed. You cannot choose not to frame when you communicate—there is no such thing as a neutral frame. Expecting resistance from scientists, Nisbet and Mooney (2007) concluded their article, "some readers may consider our proposals too Orwellian, preferring to safely stick to the facts . . . [but] as unnatural as it might feel, in many cases, scientists should strategically avoid emphasizing the technical details of science when trying to defend it" (p. 56). This quote stoked controversy because it called out a tendency to focus on the technical details of science. I want to highlight the opening charge. Rather than dismissing framing as Orwellian, I consider framing an inherent aspect of communication and language use (Bubela et al., 2009; Robbins, 2001; Nisbet & Scheufele, 2009). Even "sticking to the facts" is a particular way of framing a message—framing is inevitable.

This premise reflects my definition of framing: Framing refers to how particular ways of constructing and presenting messages results in certain impacts rather than others. This broad definition can be seen in two divergent traditions of framing. In psychology, Kahneman (2011) and colleagues have shown how human choice is contingent on how information is contextualized rather than on the expected utility of options alone. In other words, variations in how a given piece of information is presented—rather than differences in what is being communicated—can influence what people choose to do. In sociology, framing stems from Goffman (1974), who suggested that framing is fundamental to how people construct meaning and make sense of the everyday world. Frames are the interpretive schemas that people use to both classify and interpret the information they encounter in their daily lives. This moves beyond the presentations of logically equivalent information (which Kahneman focuses on) and into the territory where the selection of one set of arguments over another can be deemed a frame (Cacciatore, Scheufele, & Iyengar, 2016). How a story is told can highlight some pieces of information more than others, resulting in a particular problem definition, causal interpretation, moral evaluation, or treatment recommendation (Entman, 1993). These basic understandings of framing acknowledge that different frames can lead to messages being understood differently. But framing is not just the strategic use of message design for particular outcomes. All talk is framed. Framing is inherent in the ways that we contextualize information, determine what arguments we choose to put forward, and present data. This premise highlights the importance of developing ethics to guide framing choices since avoiding framing is not an option.

Third, the approach to framing advocated in this chapter is not the approach to framing favored by Nisbet and Mooney. Nisbet and Mooney (2007) argue that scientists should learn to actively frame issues to "make information relevant to different audiences" (p. 56). This particular approach to framing is connected to media framing research that shows people have a tendency to detect patterns in pieces of information that are consistent with preexisting cognitive schemas—how information is framed will influence the schema called upon to process that information, resulting in persuasion, priming, and agenda-setting effects (Cacciatore et al., 2016). This theory of framing fits within what I will call framing-for-persuasion. Framing-for-persuasion involves framing an issue to one's advantage in the hopes of getting an audience to do what you want it to. Framing-for-persuasion can be honest or dishonest, sincere or insincere, enlightening or manipulative. It is not inherently unethical. Nonetheless, this approach to framing is insufficient to guide science communicators committed to public engagement with science, as I'll argue below.

The shift from public understanding of science (PUS) to public engagement with science (PES) (Durant, 2008) marks a shift from presuming that mistrust toward science is generated by a knowledge gap between experts and publics to a dialogic view wherein publics are recognized as having important knowledge and experience that are valuable insights for public problem solving (Blok, Jensen, & Kaltoft, 2008). PES presumes the need for science communicators to be able to sustain a dialogue with the public on scientific controversies. Deliberative democracy—democracy grounded in individuals engaging in inclusive, respectful, and reasoned consideration of information, views, experiences, and ideas (Nabatchi, 2012; Carcasson & Sprain,

2016)—is increasingly seen as a valuable framework for engaging citizens in science and technology issues (O'Doherty & Davidson, 2010). Deliberative approaches consider citizens as equal participants along with scientific experts (Sprain, Carcasson, & Merolla, 2014); citizens provide valuable knowledge and alternative worldviews (Myskja, 2007). Deliberation provides a forum for reciprocal engagement between citizens and experts through which judgments and preferences are transformed (Petts & Brooks, 2006). The hope for deliberation as a mode of science communication is that it can provide a way of enhancing mutual understanding of science, inherent uncertainty, and value differences (Dietz, 2013).

To develop ethical principles to guide the framing done by science communicators, I consider public engagement of science and public participation on scientific, environmental, and technology issues as forms of democratic engagement. The desired means and ends of democratic engagement are defined by democratic values—"inclusiveness, participation, task sharing, lay participation, reciprocity in public problem solving, and an equality of respect for the knowledge and experience that everyone contributes to education and community building" (Hartley, Saltmarsh, & Clayton, 2010, p. 397). In turn, these democratic values can be used to evaluate the ethical basis of different forms of framing. Democratic engagement embraces the impacts that framing can have on collective decision making. Framing choices can and do impact how people understand and name public problems, perceive different actors, and cultivate opportunities for public action. Thus democratic engagement embraces framing guided by democratic values that support collective action of various kinds. When faced with a public controversy related to science, science communicators ought to consider framing as more than a means of persuasion. Science communicators should consider how framing can support democratic engagement. This chapter draws on framing-for-deliberation to develop principles and practices for framing science communication. Framing-for-deliberation uses language to clarify the range of positions surrounding an issue so that citizens can better decide what they want to do (Friedman, 2008).

The next section reviews the ethical challenges of framing, underscoring the need for ethical guidance. Then I argue why framing-for-persuasion is insufficient for democratic engagement and introduce framing-for-deliberation. I review a method for framing-for-deliberation before offering practical guidance for science communicators seeking to foster democratic engagement and engage in ethical framing.

Ethical Challenges of Framing

Framing has been developed in and applied by a broad range of academic disciplines, from sociology to political science, linguistics to communication, media studies to psychology. Rather than sort out the differences between these different traditions, this essay draws on the general concept of framing as the way that constructing and presenting messages results in certain impacts rather than others. This broad definition includes equivalence framing—framing that includes altering the presentation of logically equivalent information, such as the differences between "creation science" versus "creationism," "climate change" versus "global warming" versus "climate weirding"—and emphasis framing, which includes selecting which content to communicate to a particular audience (Cacciatore et al., 2016). Across these meanings, framing is widely acknowledged to be central to understanding how language constructs public controversies (Gamson & Modigliani, 1989).

It is easy to imagine ethically questionable framing practices. Science communicators should not be spin-doctors, using imaginative framing to deceive audiences. For example, Lakoff (2004) argues that Bush-era Republicans referred to proposals to minimize environmental regulation as being "clean," "healthy," and "safe" despite knowing that these policies would not produce these outcomes. Using deceptive language that evokes frames to garner outcomes (e.g., political support) without any concern for accuracy is deceitful and unethical. But determining what counts as deception can, in practice, be dicey. Many frame options can be tangential to science itself—focusing on economic implications when discussing climate change, for example. Ensuring accuracy is not as simple as requiring a frame to be consistent with underlying science. A more reasonable principle for determining accuracy (and in turn deception) might be that frames should not be inconsistent with underlying science. Science communicators—scientists, science journalists, politicians, and even citizens—should talk in ways that do not contradict science. But even this principle presumes a level of certainty about what counts as science not present in many public controversies (Sarewitz, 2004). Frequently scientific consensus about an emerging issue comes well after policy decisions must be made (Collins & Evans, 2008).

To illustrate the difficulty of assessing what counts as spin and deception, consider a current controversy over framing: Should extreme weather events be framed as climate change impacts? The science of attributing the cause of extreme weather and climate-related events is a growing area of research due to the practical interest in asserting relationships between weather and climate change and the difficulties of making clear attributions. Weather events happen on a wide range of timescales. A weather extreme such as a very high daily rainfall total is harder to connect to anthropogenic climate change than a climate-relate extreme, such as a very high seasonal mean temperature (Stott et al., 2016). There is a basic expectation that climate change will alter the occurrence of some extremes, but attribution for a particular event relies on knowing how the probability of the event is changing and whether or not anthropogenic climate change has altered the probability or the magnitude of a particular event. Individual studies show clear evidence for human influence on some weather events and little evidence for human influence on others. Parsing the differences often involves consideration of both natural variability and how the probability of the condition itself is changing. This can reveal unexpected interactions, such as climate change reducing the probability for 2013 Colorado floods since the probability of rainfall had likely decreased due to climate change as a result of changes in atmospheric circulation and vertical stability.

When, then, is it trickery to frame extreme weather events as climate change impacts? A newspaper article could, for example, say that climate change will create more weather problems, including more storms and floods. This would be considered a valid science frame (Antilla, 2005). When framing shifts from a general argument that climate change may result in future extreme weather events to linking a specific event with climate change or arguing that current weather patterns were caused by climate change, things get dicey. Nonetheless, several advocacy groups favor and encourage this framing since it can help connect local weather with climate in ways that make climate change impacts tangible to the public. The advocacy group 350.org held a day of action in 2012 designed to help "connect the dots" between extreme weather and climate change. They encouraged people to take pictures of damage from hurricanes, tornados, and more with large dots labeling the wreckage as climate change. Select media coverage over the past decade has also suggested that extreme weather events are caused by climate change (Carvalho & Burgess, 2005; Antilla, 2005).

Some scientists and science communication practitioners have challenged these frames, arguing that there is seldom empirical proof that specific extreme weather events have been caused by climate change (e.g. Bouwer, 2011; Kloor, 2012). Yet, based on a review of the literature on weather and climate change, Trenberth (2012) argues that climate change has become so significant that all weather events are impacted by climate change. Attempting to moderate this controversy, the Union of Concerned Scientists (Ekwurzel, 2012) developed informational graphics that placed extreme weather events on a continuum, ranging from limited evidence of linking with human-caused climate change (tornados and hurricanes) to strong evidence of linking (coastal flooding and heat waves). Nonetheless, debates have persisted over what links can and should be made between extreme weather events and climate change.

In this example, what counts as spin? Is it unethical (or inaccurate) to invoke climate change in discussions about Hurricane Sandy? Is it unethical *not* to invoke climate change when discussing this disaster? Is it okay to imply that the wreckage from coastal flooding or a forest fire in the Western United States should motivate citizens to work toward curbing climate change? Is framing all extreme weather events as related to climate change ethical even though attributing any particular extreme weather event to climate change is questionable?

This framing controversy reveals two important lessons for the development of the ethics of framing. First, determining what counts as a legitimate scientific basis to justify a particular frame may be difficult to establish. This is particularly true when we begin to parse the differences between causation and correlations, between projected trends and current pathways, between general knowledge and application to a specific case. In the case of climate modeling, science is based on probabilistic modeling so it will only be able to show correlations, not causation, and the model is likely focused on general trends not specific cases. Second, this example is about the *accuracy* of framing extreme weather events as climate change impacts, but it is also over the *efficacy* of using frames to achieve particular communication goals. Activists want to connect extreme weather with climate change because they believe that these connections will motivate people to push for political action on climate change—this goal is not unethical. After all, climate change is notoriously difficult for individuals to see and experience. Individuals can see and experience extreme weather. Thus extreme weather events provide a prime opportunity for overcoming one of the fundamental challenges for communicating about climate change. For some communicators, the ultimate goal—mobilizing political action—warrants rhetorical use of extreme weather events. On the other side, scientists and practitioners are concerned that misrepresenting science will undermine the credibility of arguments for climate change, confuse those who make decisions, and lead to poor decision making. I am not taking a stance against advocacy. Instead, I want to distinguish communication goals from framing itself. An ethics of framing should be able to take into consideration what a communicator is trying to accomplish through frame selection.

Framing raises ethical questions for science communication. These questions cannot be resolved by avoiding framing—there is no such thing as a neutral frame or an unframed message. Instead, framing obligates us to reflect on what we are trying to accomplish with language, including how our choices may have unintended consequences. The next section argues that framing-for-persuasion can, sometimes, work counter to democratic engagement and introduces alternative principles to guide framing by science communicators seeking to support public engagement with science through framing-for-deliberation.

Framing for Democratic Engagement

Framing in science communication research is often associated with studying how particular frames resonate with particular audiences (e.g., Anderson, 2009; Nisbet & Scheufele, 2009). The effectiveness of a particular frame is based, in part, on its relevance; thus science communication scholars will research the relevance of particular frames for specific audiences. Working from this premise, studies in science communication often evaluate how the presentation of an issue can produce changes of opinion (Chong & Druckman, 2007), such as how framing climate change in terms of economic benefits (Leiserowitz, 2006), health concerns (Maibach, Nisbet, Baldwin, Akerlof, & Diao, 2010), or stewardship and religious values (Zia & Todd, 2010) appeal to particular audiences. Within this paradigm, the goal of frame analysis can be to maximize persuasion and motivate greater interest in science issues (Nisbet & Scheufele, 2009). This is what I call framing-for-persuasion (Friedman, 2008). Here I use persuasion in the broad sense of the term, including language use to produce changes of opinion, language use to make information relevant to different audiences, and language use to mobilize people to advocate for particular policy actions. In all of these scenarios, aligning frames with existing worldviews is a key element in nudging the public to support policies informed by science or tackle scientific controversies in particular ways.

Within a vibrant public sphere, science communicators have many occasions when persuasion is necessary and appropriate for a variety of political and scientific goals. I don't consider it inherently unethical. Nonetheless, framing-for-persuasion alone is insufficient for enabling democratic engagement on science and technology public controversies—indeed, it may undermine this goal. When transparent, framing-for-persuasion can undermine trust with publics if people feel as though they are being manipulated or pressured into thinking a particular way (Bubela et al., 2009). Instead of seeing scientists as honest brokers (Pielke, 2007) who help lay out options, framing-for-persuasion can mark science communicators as issue advocates. Attempting to accomplish change through "effective messages" that influence public opinion can also result in supporting short-term pragmatic actions that fit within economic and political imperatives but fail to consider broader ecological systems and public values (Brulle, 2010).

Moreover, using frames that resonate with particular audiences can lead to increased audience segmentation and messaging to distinct audiences. The more that frames attempt to resonate with existing interpretative schema, the more that these frames may end up feeding political polarization and disagreement by associating science controversies with other politicized issues. Instead of simplification, democracy requires citizens to be aware of differences. An important illustration comes from how a public problem is framed (e.g., what is the public problem and how severe is it) (Juntti, Russel, & Turnpenny, 2009). Framing-as-persuasion can focus on promoting a single definition of the problem: Climate change is a religious issue or a health issue or an economic issue or an environmental issue. Yet this approach does not foster democratic engagement that can work across and between different perspectives, particularly when different stakeholder groups have different understandings of the fundamental problem. Instead, fostering public action often requires overtly juxtaposing and negotiating different problem definitions rather than masking conflicts by pushing a single, narrow frame. Addressing public controversies often requires collective action that reaches across political and ideological divides; audience segmentation may, unintentionally, perpetuate these divides by developing different, narrow ways of framing the problem and acceptable solutions.

Fostering public engagement requires more than effective messages that favorably influence public opinion (Brulle, 2010). Making progress on controversial environmental and scientific issues requires the opposite approach: overtly juxtaposing and dealing with conflicting discourses and ways of framing environmental problems (Juntti et al., 2009). If, as Nisbet and Scheufele (2009) argue, the future of science communication is facilitating conversations with the public that recognize, respect, and incorporate differences in knowledge, values, perspectives, and goals, framing strategies must enable the public to recognize and reconcile differences rather than just work within existing schema.

These goals can be better accomplished by drawing on framing-for-deliberation. Deliberation foregrounds the importance of people coming together to consider an issue from multiple perspectives to weigh the tradeoffs between different perspectives under conditions of respect and mutual consideration (Carcasson & Sprain, 2012, 2016). Framing-for-deliberation means clarifying the range of positions surrounding an issue so that citizens can better decide what they want to do (Friedman, 2008). Framing-for-deliberation often involves exposing a group to multiple frames for understanding a particular issue. For example, consider three approaches to addressing climate change offered by a Public Agenda guide for citizen thought and action:

(1) We need decisive local, national, and interactional action to prevent and minimize the worst consequences of climate change.

(2) We need to make sure our most vulnerable communities adapt to the inevitable changes global warming will cause.

(3) We should trust the free market to lead the way in the search for solutions.

Each of these approaches frames climate change differently (e.g., preventable, inevitable, a general problem) and suggests distinct strategies for addressing it (e.g., mitigation, adaptation, technology). In framing-for-deliberation, participants would be exposed to not one but all three different ways of addressing climate change. These approaches are not mutually exclusive. Instead, framing-for-deliberation suggests that a full conversation would consider all of them—even if any specific community decided to combine them or innovate new solutions beyond these frames. Practically, this is equivalent to exposing participants to multiple contradictory frames. Empirical research suggests that some persuasive framing effects disappear when individuals understand the rationales for multiple frames (Druckman & Nelson, 2003). Framing-for-deliberation intentionally uses crosscutting interpersonal discussions and background materials to limit the framing effects of a single message and, instead, encourage individuals to come to considered judgment about an issue by making them reflect on multiple, competing frames.

Principles and Practices for Framing for Democratic Engagement

Deliberative work includes political theory, empirical research, a movement for political reform, and a profession of forum design and facilitation (Dryzek, 2002). Deliberative practitioners who work with US organizations such as the National Issues Forum or the Kettering Foundation's Centers for Public Life have been framing-for-deliberation for over twenty years. They develop issue guides and organize deliberative forums. Their experience provides some basic practices to guide framing for democratic engagement through framing-for-deliberation. In this context, framing refers to the communication processes structuring deliberation or democratic discussion, including definition and construction of the issue under deliberation, development of alternatives at stake, emphasizing some elements at the expense of others, and suggesting interpretive connections among certain ideas and symbols (Barisione, 2012).

Generating these deliberative frames includes three basic steps. First, framing-for-deliberation starts with broad issue analysis in order to understand the nature of the problem (Carcasson & Sprain, 2016). Fundamentally, democratic action relies on some shared understanding of the problem in order to provide traction for different people to come together and act. If people disagree about the nature of the problem (or do not recognize an issue as a problem), they often will struggle to come together to address the problem. In the United States, climate change discussion has suffered from this issue as people contest whether climate change is happening or whether it is human-caused rather than agreeing climatic changes are fundamentally a public problem. Within framing-for-persuasion, communicators often generate new frames that each construct a distinct diagnosis of the problem. Instead, framing-for-deliberation attempts to define a problem or constellation of problems in a way that broadly resonates with a community so the problem framing is widely accepted, and this problem can be considered from multiple different perspectives. The problem should be stated in terms that take into account things that people hold valuable rather than a technical diagnosis of the issue (e.g., too many children left behind vs. the achievement gap) (Rourke, 2015).

Framing-for-deliberation presumes that political issues are public in nature and require collective action. This means that issues are framed as public issues, not technical issues. Yet it attempts to avoid treating issues as "merely political," denigrating politics to the messy business of tracking the whim of the public. As Moore (2010) observes, framing issues such that some kinds of concerns appear legitimate while others are merely political or transient matters of public concern can trivialize problems and lead to them being dismissed. When considering framing-for-deliberation, science communicators should interrogate whether a particular issue requires public action. This is, of course, not a foregone conclusion. Many science communicators have long considered the differences between technical sphere issues and public issues (e.g., Collins & Evans, 2002; Goodnight, 2012). Some issues should be framed as technical issues and left to the appropriate experts. Science communicators alone are not responsible for determining whether an issue is technical or public. As I argue below, framing for democracy opens up framing to public challenging, which includes the public arguing that issues are public (as opposed to experts alone making this determination). But framing-for-deliberation supports public framing that enables meaningful public engagement with the issue.

Next, framing for democratic engagement attempts to disentangle key elements of a complex problem in such a way that people from a wide variety of backgrounds and starting points are able to grapple with the shared problem or constellation of problems. The goal here is to provide sufficient detail for citizens to engage an issue without leaving them completely overwhelmed. Citizens do not need to be "secondhand scientists" in order to deliberate about an issue (Kadlec, 2009). Instead, they need to be given key information that helps them get past common misconceptions and a range of choices that are presented in nontechnical language so that they can weigh the costs and trade-offs of possible approaches. The Kettering Foundation argues that citizens should have enough factual material and data so that citizens from different circumstances who deliberate together have the knowledge they need to engage productively, focusing on data that illuminate the important conflicts or dilemmas that the issue raises (Rourke, 2015).

Finally, there is a general move away from binary framing, presenting an issue as all or nothing, us versus them, or a single yes or no decision. Often, within a democracy, issues are framed in terms of voting on a particular proposition. Initiatives, for example, have citizens vote yes or no on a proposed law; in our two-party system, there is a left and a right position. This issue framing generates advocates on both sides of an issue. Framing-for-deliberation attempts to resist having two sides to an issue. Instead, issues are explored from multiple perspectives, including how various stakeholders might approach a particular topic. A simple way of avoiding binary thinking is to include more than two positions to an issue. Many deliberative framings have three to four distinct approaches that each have multiple potential actions within them. The purpose is not to simply vote one approach as the best, but instead to fully explore the issue. Within these distinct approaches, trade-offs highlight the consequences of different actions so that people can wrestle with the implications of any single approach.

Framing Science

Some science communicators may commit to developing an entire framework for deliberation similar to the approach described above. For example, the North American Association for Environmental Education is partnering with the Kettering Foundation to develop issue guides on environmental topics such as water scarcity and climate change that can be used by their network of environmental educators. Yet many science communicators will stop well short of developing entire frameworks. Nonetheless, insights from framing-for-deliberation can provide a set of principles and practices for framing science.

First, citizens must be able to challenge framing, including assumptions about science. In her analysis of British Columbia's deliberation on bio-banking, Walmsley (2009) notes that citizens challenged many of the fundamental assumptions held by the organizers who framed the event. Participants developed and embellished the figure of a "mad scientist" as a way to challenge the certainties promised by scientific, legal, and ethical expertise within the event. They questioned whether science can be governed and challenged the assumption that local governance is of any use at all. Moreover, they challenged the assumption that citizens are interested in democratizing science. All of these challenges are legitimate and encouraged within democratic engagement. Indeed, they uphold the democratic values of lay participation, reciprocity in public problem solving, and an equality of respect for the knowledge and experience that everyone contributes to education and community building. Thus ethical science communication framing should be open to challenge and reframing by citizens.

Second, science communicators should ask: how much factual material does the public need to engage this issue productively? Within framing-for-deliberation, factual material does not resolve public controversies; instead, it supports the exploration and consideration of disagreement that is the essence of democracy. Science communicators aiming to support public engagement of science should consider what science the public needs to do this work of citizens. This requires asking: does this information illuminate what is at issue—the important conflicts or dilemmas? These questions shift science communicators to thinking about frame information for citizens, not scientists (rather than attempting to turn the public into experts). These questions have implications for how science communicators frame messages, including the rhetorical forms they use (e.g., using stories, graphics, statistical results) as well as the overall complexity, nuance, and amount of information provided. This does not, of course, mean that science communicators get to determine the only relevant information that the public should have access to; framing-for-deliberation is not about engineering a narrow set of valid outcomes or options for actions. Instead, it means a shift in thinking about how to evaluate what the public needs to know.

Third, when there are conflicting ways of defining the problem, the underlying science, or approaches to the issue, science communicators should aim to present and juxtapose multiple options rather than present a singular story. This might mean museum displays that lay out different ways that people think about a controversial issue like how to supply energy in the future. Journalists can move beyond the tendency to tell two sides of an issue to instead reveal a broader array of approaches to an issue within a single community. Determining the situations that call for this will, of course, be a matter of judgment. Framing for democratic engagement does not reject the possibility of scientific consensus nor does it aim to perpetuate manufacturing controversies (see Ceccarelli, 2011). Democratic values can be misused to invent controversy where it does not exist; framing-for-persuasion can wrongly minimize controversy where it does exist. Ethical framing of science calls for critical reflection about the individual circumstances. When in doubt, framing for democratic engagement favors juxtaposing multiple perspectives so that citizens can distinguish between them.

Conclusion

Rather than focus on developing principles and practices for adjudicating spin and misinformation, framing for democratic engagement orients science communicators to the rhetorical situation of public engagement. Democracy requires public participation to wrestle with controversial issues related to science, technology, and environmental issues. Science communicators supporting public engagement with science should aim to build reciprocal relationships that foster an equality of respect for the knowledge and experience that everyone contributes to addressing public problems. Nonetheless, publics need enough knowledge to productively engage in deliberation and discussion about public issues. Science communicators—journalists, environmental educators, museum curators, extension agents, and willing scientists—are called on to provide these essential means to facilitate public engagement and public action on pressing scientific, environmental, and technological issues.

References

Anderson, A. (2009). Media, politics and climate change: Towards a new research agenda. *Sociology Compass, 3*, 166–182.

Antilla, L. (2005). Climate of skepticism: US newspaper coverage of the science of climate change. *Global Environmental Change, 15*, 338–352.

Barisione, M. (2012). Framing a deliberation: Deliberative democracy and the challenge of framing processes. *Journal of Public Deliberation, 8*, article 2. Retrieved from http://www.publicdeliberation.net/jpd/vol8/iss1/art2

Blok, A., Jensen, M., & Kaltoft, P. (2008). Social identities and risk: Expert and lay imaginations on pesticide use. *Public Understanding of Science, 17*, 189–209.

Bouwer, L. M. (2011). Have disaster losses increased due to anthropogenic climate change?. *Bulletin of the American Meteorological Society, 92*(1), 39–46.

Brulle, R. J. (2010). From environmental campaigns to advancing the public dialog: Environmental communication for civic engagement. *Environmental Communication, 4*, 82–98.

Bubela, T., Nisbet, M. C., Borchelt, R., Brunger, F. Critchley, C., Einsiedel, E., . . . Caulfield, T. (2009). Science communication reconsidered. *Nature Biotechnology, 27*, 514–519.

Cacciatore, M. A., Scheufele, D. A., & Iyengar, S. (2016). The end of framing as we know it . . . and the future of media effects. *Mass Communication and Society, 19*, 7–23.

Carcasson, M., & Sprain, L. (2012). Deliberative democracy and adult civic education. *New Directions in Adult Civic Education, 135*, 15–23.

Carcasson, M., & Sprain, L. (2016). Beyond problem solving: Re-conceptualizing the work of public deliberation as deliberative inquiry. *Communication Theory, 26*, 41–63.

Carvalho, A., & Burgess, J. (2005). Cultural circuits of climate change in U. K. broadsheet newspapers, 1985–2003. *Risk Analysis, 25*, 1457–1469.

Ceccarelli, L. (2011). Manufactured scientific controversy: Science, rhetoric, and public debate. *Rhetoric & Public Affairs, 14*(2), 195–228.

Chong, D., & Druckman, J. N. (2007). Framing theory. *Annual Review of Political Science, 10*, 103–126.

Collins, H., & Evans, R. (2008). *Rethinking expertise.* Chicago, IL: University of Chicago Press.

Collins, H. M., & Evans, R. (2002). The third wave of science studies: Studies of expertise and experience. *Social Studies of Science, 32*(2), 235–296.

Dietz, T. (2013). Bringing values and deliberation to science communication. *Proceedings of the National Academy of Sciences, 110*, 14081–14087.

Dryzek, J. S. (2002). *Deliberative democracy and beyond: Liberals, critics, contestations.* New York, NY: Oxford University Press.

Druckman, J. N., & Nelson, K. R. (2003). Framing and deliberation: how citizens' conversations limit elite influence. *American Journal of Political Science, 47*, 729–745.

Durant, D. (2008). Accounting for expertise: Wynne and the autonomy of the lay public actor. *Public Understanding of Science, 17*, 5–20.

Ekwurzel, B. (2012, July 16). Evidence check: Which extreme weather events are more linked with climate change—heat waves or hurricanes? [Weblog post]. *The Equation: A blog on independent science and practical solutions.* Retrieved from http://blog.ucsusa.org/extreme -weather-and-climate-change

Entman, R. M. (1993). Framing: Toward clarification of a fractured paradigm. *Journal of Communication, 43*(4), 51–58.

Friedman, W. (2008). *Reframing "framing."* Occasional Paper, No. 1, Center for Advances in Public Engagement. Retrieved from http://www.publicagenda.org/files/Reframing %20Framing.pdf

Gamson, W. A., & Modigliani, A. (1989). Media discourse and public opinion on nuclear power: A constructionist approach. *American Journal of Sociology, 95*, 1–37.

Goffman, E. (1974). *Frame analysis: An essay on the organization of experience.* Cambridge, MA: Harvard University Press.

Goodnight, G. T. (2012). The personal, technical, and public spheres: A note on 21st century critical communication inquiry. *Argumentation and Advocacy, 48*(4), 258–268.

Hartley, M., Saltmarsh, J., & Clayton, P. (2010). Is the civic engagement movement changing higher education? *British Journal of Educational Studies, 58*(4), 391–406.

Juntti, M., Russel, D., & Turnpenny, J. (2009). Evidence, politics and power in public policy for the environment. *Environmental Science & Policy, 12*, 207–215.

Kadlec, A. (2009). Mind the gap: Science museums as sources of civic innovation. *Museums & Social Issues, 4*, 37–53.

Kahneman, D. (2011). *Thinking, fast and slow.* New York, NY: Farrar, Strauss, Giroux.

Kloor, K. (2012, September 21). The search for a winning climate change frame [Weblog post]. Retrieved from http://blogs.discovermagazine.com/collideascape/2012/09/21/the-search-for-a-winning-climate-change-frame/

Lakoff, G. (2004). *Don't think of an elephant: Know your values and frame the debate.* New York, NY: Chelsea Green.

Leiserowitz, A. (2006). Climate change risk perception and policy preferences: The role of affect, imagery, and values. *Climatic Change, 77,* 45–72.

Maibach, E. W., Nisbet, M., Baldwin, P., Akerlof, K., & Diao, G. (2010). Reframing climate change as a public health issue: An exploratory study of public reactions. *BMC Public Health, 10,* 299.

Moore, A. (2010). Public bioethics and public engagement: the politics of "proper talk." *Public Understanding of Science, 19*(2), 197–211.

Myskja, B. K. (2007). Lay expertise: Why involve the public in biobank governance? *Genomics, Society, and Policy, 3,* 1–16.

Nabatchi, T. (2012). An introduction to deliberative civic engagement. In T. Nabatchi, J. Gastil, G. M. Weiksner, & M. Leighninger (Eds.), *Democracy in motion* (pp. 3–18). New York, NY: Oxford University Press.

Nisbet, M. C. (2007). Framing Science sparks a seismic blog debate [Weblog post]. Retrieved from http://bigthink.com/age-of-engagement/framing-science-sparks-a-seismic-blog-debate

Nisbet, M. C. (2009). The ethics of framing science. In B. Nerlich, R. Elliott, and B. Larson (Eds.), *Communicating biological sciences: Ethical and metaphorical dimensions* (pp. 51–73). London, UK: Ashgate.

Nisbet, M. C., & Mooney, C. (2007). Framing science. *Science, 316,* 56.

Nisbet, M. C., & Scheufele, D. A. (2009). What's next for science communication? Promising directions and lingering distractions. *American Journal of Botany, 96,* 1767–1778.

O'Doherty, K. C., & Davidson, H. J. (2010). Subject positioning and deliberative democracy: Understanding social processes underlying deliberation. *Journal for the Theory of Social Behavior, 40,* 224–246.

Petts, J., & Brooks, C. (2006). Expert conceptualizations of the role of lay knowledge in environmental decisionmaking: Challenges for deliberative democracy. *Environment and Planning A, 38,* 1045–1059.

Pielke, R. A. (2007). *The honest broker: making sense of science in policy and politics.* Cambridge, UK: Cambridge University Press.

Robbins, P. (2001). Fixed categories in a portable landscape: the causes and consequences of land-cover categorization. *Environment and Planning A, 33,* 161–179.

Rourke, B. (2015). *Developing materials for deliberative forums.* Dayton, OH: Kettering Foundation.

Sarewitz, D. (2004). How science makes environmental controversies worse. *Environmental Science & Policy, 7,* 385–403.

Sprain, L., Carcasson, M., & Merolla, A. (2014). Experts in public deliberation: Lessons from a deliberative design on water needs. *Journal of Applied Communication Research, 42,* 150–167.

Stott, P. A., Christidis, N., Otto, F. E. L., Sun, Y., Vanderlinden, J., van Oldenborgh, . . . Zwiers, F. W. (2016). Attribution of extreme weather and climate-related events. *WIREs Climate Change, 7,* 23–41.

Trenberth, K. E. (2012). Framing the way to relate climate extremes to climate change. *Climatic Change, 115*(2), 283–290.

Walmsley, H. L. (2009). Mad scientists bend the frame of biobank governance in British Columbia. *Journal of Public Deliberation, 5*, 1–26. Retrieved from http://www.publicdeliberation.net/jpd/vol5/iss1/art6

Zia, A., & Todd, A. M. (2010). Evaluating the effects of ideology on public understanding of climate change science: How to improve communication across ideological divides? *Public Understanding of Science, 19,* 743–761.

PART TWO

Professional Practice

Good communication is always adapted to its context. How something ought to be said varies depending on the audience, the communicator, the medium, the actual message, previous communication, and a host of other factors. Whereas the first part of this book explored ethical perspectives on science communication in general, the chapters in this part focus on a small sample of the diverse ethical challenges arising in particular contexts.

Science communicators are often advised not to communicate science to publics using the same techniques that scientists use with each other. Dahlstrom and Ho (chapter 5) open this part with an examination of the ethics of one of the possible alternative techniques. In many ways, storytelling can offset some of the flaws typical in traditional science presentations. The social science literature summarized in the essay shows that narrative engages ordinary, not specialist, styles of reasoning; a story is familiar, vivid, and memorable instead of unfamiliar, abstract, and confusing. But this means that narrative can persuade covertly and without giving reasons, making it a powerful, and thus also dangerous, communication technique. For example, one compelling anecdote about vaccine harm can overwhelm a pile of statistical evidence of vaccine safety. In any particular context, should narrative be used at all for communicating science? If so, what goals can it accomplish, and to what extent is it appropriate to fictionalize details, increasing vividness at the possible cost of accuracy?

Since the development of the art of rhetoric in ancient Greece, communication theorists have recognized that success depends on having a compelling, well-supported message (logos), on putting the audience in the right frame of mind to receive that message (pathos), and on showing the messenger to be a person of good character, good will, and good judgment (ethos).

Ranalli (chapter 6) focuses on this last aspect, what he calls the "personal dimension of science communication." As he points out, even the impersonal prose scientists use to address each other conveys the character of the writer. This is best understood not as a "fraud" or fake performance, but as a form of self-discipline; scientists in addressing other scientists conspicuously strive to live up to ideals of disinterestedness, lack of bias, independence, truthfulness, and objectivity. And the public expects the same when scientists turn to address them. In addition to considering the quality of the science directly, Ranalli suggests that audiences judge the characters of the persons presenting the science, deciding whether or not to trust them.

The last three chapters in this part consider the ethical challenges facing three distinct types of science communicators: the journalist, the industry public relations professional, and the scientist him- or herself. The first is perhaps the first to come to mind, since even in this networked era science journalists continue to play key roles in supporting the circulation of scientific results to broader publics. On the basis of a study of the use of expert sources in news reports on bioethics, Kruvand (chapter 7) examines the tensions journalists experience when relying on technical experts. Constraints including short deadlines and lack of expertise force journalists to rely on experts when reporting on technical subjects. But few experts are able to work effectively with the press, in part because it asks them to communicate in ways that go against their training, such as by simplified, colorful, rapidly produced sound bites. The unfortunate result: Just a few experts tend to dominate national conversations.

Then Gaither and Sinclair (chapter 8) explore the use of environmental claims in industry advertising. Some readers may be surprised by the idea that advertising has an ethics at all. But it is important to realize that unethical behavior by some communicators undermines the trustworthiness of publicity for all, so businesses that are honestly committed to the public good have a strong interest in preserving a clean marketplace of ideas. Ethical PR intersects with science communication in industry environmental campaigns, which portray industry initiatives as having positive impacts on climate change, pollution, biodiversity, and other legitimate and important issues. Such campaigns are often intended to build relationships with consumers in the economic realm and sometimes to push public opinion against regulation in the political realm, so their self-interested motive is clear. What ethical standards do the campaigns need to live up to, in order to be not only effective in achieving business aims but also legitimate and credible?

This part closes by taking up a central question: Do scientists have a responsibility to communicate with the public at all? Davies (chapter 9) traces the history of the society-wide conversation that has come to recognize public engagement as a duty. But what do scientists themselves think? Davies' interviews with 29 research team leaders show that science faculty members are often deeply committed to ethical ideals, seeing their lives as embedded in a complex network of responsibilities. Scientists understand that they owe to science itself a responsibility to do good work—work that adds something to their fields. They also owe to the students and staff in their labs a duty of care for their professional and personal development. And they consider themselves accountable to the public to make sure that their work has an impact. Problems arise when these responsibilities are in tension, however—including tensions caused by the most basic problem of limited time. In such cases, immediate responsibilities to their own work and that of their labs may win out over an abstract obligation to "the public."

The central theme emerging in this part seems to be that ethical professional practice in science communication can run up against a variety of barriers and challenges: whether a more compelling format like narrative could mislead audiences more than inform them, whether those audiences will feel they can trust the character of the scientist-communicator, whether time and resources allow inclusion of a wider range of source voices in news reports, whether corporate environmental campaigns can meet business goals without violating ethical principles, and whether it is even realistic to expect scientists to embrace a new ethical responsibility (public engagement) alongside the ones they already have. Perhaps we need to expand the conversation beyond "scientists should engage" and "corporations should be more transparent" to a question for all science communicators: How can we best balance our many responsibilities to society with our responsibilities to our professions and to the organizations that we represent?

5

Exploring the Ethics of Using Narratives to Communicate in Science Policy Contexts

MICHAEL F. DAHLSTROM
AND SHIRLEY S. HO

There is a growing sense that scientific information is not contributing what it should to controversial science policy. Social controversies surrounding topics such as climate change, evolution, and vaccination are often claimed to exemplify either an ignorance of scientific data or its outright rejection (Baker, 2008; Forrest, 2001; Mooney, 2005; Zimmer, 2011). Science can never instruct society in what it should do, as personal and collective ethics define what a society values (Volti, 2009). However, confusion and mistrust of science undercut the foundation on which these collective values can best be achieved.

This perceived lack of scientific influence within policy making has not gone unnoticed. National calls have described the need for more effective science within policy making (Basken, 2009), and initiatives such as the American Association for the Advancement of Science (AAAS) "Communicating Science" program have responded by offering researchers techniques to increase the clarity of their communication (AAAS, 2016). Likewise, various strands of research have explored techniques for improving the effectiveness of science communication within a policy context, including framing (Durfee, 2006), trust-building (Liu & Priest, 2009), and altering the top-down communicative model in which science communication is often conceived (Dickson, 2001; Nisbet, 2009).

Another technique relevant to the communication of science is narrative. Narrative describes a format of communication involving a temporal sequence

This chapter has been updated and modified from an article previously published in *Science Communication: Linking Theory and Practice*, 34(5): 592–617, October 2012.

of events influenced by the actions of specific characters. Examples of narrative range from short exemplars or testimonials that may be contained within larger messages to detailed and lengthy entertainment stories common in the movie and book industries. Research suggests that narrative communication is encoded using a unique cognitive pathway and results in effects that are quite different from argumentative or evidence-based communication. Specifically, narrative communication often improves comprehension (Graesser, Olde, & Klettke, 2002), generates more interest and engagement with a topic (Green, 2004, 2006; Green & Brock, 2000), increases self-efficacy through modeling (Oatley, 1999; Slater & Rouner, 2002), influences real-world beliefs (Dahlstrom, 2010, 2015; Slater, Rouner, & Long, 2006), and can be more successful for persuading an otherwise resistant audience (Moyer-Guse & Nabi, 2010). As such, narratives hold promise for improving the effectiveness of science communication to nonscientist audiences (Dahlstrom, 2014) and have been examined with regard to science-related topics such as health (Hinyard & Kreuter, 2007; Winterbottom, Bekker, Conner, & Mooney, 2008), risk (de Wit, Das, & Vet, 2008; Golding, Krimsky, & Plough, 1992), and the environment (Dahlstrom, 2010, 2012).

However, while much research has focused on the effects of narrative communication in a science policy context, little has examined the ethical considerations of doing so. Narrative communication may offer benefits toward a particular set of communicative goals, but what ethical considerations exist at the intersection of narrative influence and the role of science within society? We will address this gap in the literature by exploring the ethical considerations scientists and science communicators face when considering a narrative strategy for their communicative goals. To do so, we will first highlight the role of narrative and its relation to science communication. Second, we will define the scope of our discussion within the larger conversation of ethics in science communication. Finally, we will introduce three ethical considerations faced by science communicators when using narrative in a science policy setting. These considerations intersect with perceptions of the appropriate roles of communication and of scientists within democracy.

Narrative in Science Communication

The relationship between science and narrative is often discussed in one of two conflicting contexts. The first context sets up a dichotomy between science and narrative based on differences in cognitive processes that underlie comprehension. Much of the literature underlying this context comes from discourse and cognitive psychology, which explores how the mind comprehends narrative information compared to other types of information. The second context treats narrative as a communicative technique able to enhance the persuasive impact of scientific information. Much of the literature underlying this context comes from the field of narrative persuasion, which explores the often-covert influence of narrative on beliefs and attitudes. We will discuss each context in turn.

Narratives play an influential role in how individuals comprehend the world. At a cultural level, the concept of narrative has a close relationship with that of frames and metaphors in that they all organize perception through their symbolic power, their ability to relate beliefs, values, and actions, and their widespread recognition within society (Hertog & McLeod, 2001). At a cognitive level, narratives have been said to represent the default format for human thought, which forms the foundation for decision making (Schank & Abelson, 1995). This reliance on narratives is suggested to be an evolutionary response to the need of humans to model the thoughts of other humans in the complex social interactions that define our species (Read & Miller, 1995).

Research into narrative crosses diverse disciplines (Kreiswirth, 1992), resulting in a confusing array of narrative conceptualizations. For the purposes of this chapter, we define narrative as a format of communication involving a temporal sequence of events influenced by the actions of specific characters, noting that the most frequent format in our science-policy context will be that of text or speech. Narrative defined in this manner is often contrasted with other formats of communication (Longacre, 1983), most notably with that of the evidence-based communication underlying science.

Such contrasts have led to a proposed division between a paradigmatic and narrative pathway of cognition (Bruner, 1986; Fisher, 1984). The paradigmatic pathway controls the encoding of evidence-based arguments while the narrative pathway controls the encoding of situation-based exemplars. Not only does research support this split, but it also suggests the balance is not equal. Narrative text is recalled twice as well and read twice as fast as evidence-based text (Graesser et al., 2002) and generates greater engagement, persistent attitude and belief changes, and self-efficacy (Appel & Richter, 2007; Green & Brock, 2000; Oatley, 1999; Slater & Rouner, 2002). Many of these benefits are due to narrative's cause-and-effect structure. Causal relations have been shown to drive much of narrative processing (Dahlstrom, 2010, 2012; Graesser et al., 2002) and the perception of events changing over time both provides a mental simulation of how the world works (Oatley, 1999) and serves to limit the possibility of future choices, making the resolution of the narrative seem inevitable (Curtis, 1994).

A tangible consequence of the difference between narrative and para-digmatic pathways within science communication can be illustrated by the controversy between childhood vaccines and autism. Scientific studies have repeatedly found no link between the two, but a significant group of parents and celebrities have mobilized under a convincing narrative that someone's child developed autism just after receiving the vaccination. This represents a strong evidence-based argument in direct conflict with a strong narrative and the results are often troubling to scientists. When asked about the growth of the anti-vaccine movement in a *New York Times* article from June 2005, Dr. Melinda Wharton, deputy director of the National Immunization Pro-gram, stated, "This is like nothing I've ever seen before. . . . It's an era where it appears that science isn't enough" (Harris & O'Conner, 2005). Rather than a diminishing faith in science, it is probable that both sides are processing the same information differently, and the misunderstanding between paradig-matic and narrative processing may be perpetuating the conflict.

The second context relating science and narrative is not one of contrast, but of cooperation. In this context, science can use narrative to achieve its communicative goals by unobtrusively changing perceptions about the world through narrative's ability to create meaning with a veiled normative compo-nent (Bruner, 1986, 1991; Fisher, 1984).

Narratives imply a strong normative assessment of thought and action yet neither state nor defend the assumptions upon which they rely (Bruner, 1991). Because "what makes a good story is different from what . . . makes it true" (Mink, 1978, pp. 129–130), incorrect narratives are rarely influenced by evidence, and instead require a more convincing narrative to counter (Kreiswirth, 1992). The fact that narratives are able to construct reality and provide values to real-world objects without argument makes it difficult to counter their claims, and the ease with which they are processed amplifies their influence.

The field of narrative persuasion examines how communication prac-titioners can take advantage of narrative comprehension to overcome re-sistance to their messages. Studies often expose a participant to a narrative and then afterward measure whether the participant accepted the normative view of the narrative or the specific facts mentioned within, often contrasted with a non-narrative or statistical treatment. Results suggest that individu-als are often more willing to accept normative evaluations from narratives than evidence-based arguments (Braddock & Dillard, 2016; Green & Brock, 2000; Slater & Rouner, 2002). A common barrier to traditional persuasion is the formation of counterarguments that block the acceptance of a persuasive message. However, persuasive narratives have been found to reduce the formation of counterarguments (Dal Cin, Zanna, & Fong, 2004; Green, 2006; Green & Brock, 2000). Acceptance of narrative evaluations has, therefore, been described as a default outcome of exposure, where rejection is possible only with added scrutiny after the fact (Gerrig, 1993; Green, 2006).

It could be argued that fictional entertainment narratives should be discounted, or at least granted lesser weight than truthful narratives. However, studies generally find that individuals use information from fictional stories to answer general questions about the world (Appel & Richter, 2007; Dahlstrom, 2010, 2012, 2015; Marsh, Meade, & Roediger, 2003), and manipulation checks to ensure that participants understood the story was not true show that individuals do not discredit information just because a narrative is labeled fictional (Green & Brock, 2000). Even when individuals perceive much of the information in a narrative to be inaccurate, the narrative is rarely rejected completely (Appel & Richter, 2007; Marsh et al., 2003).

Narrative persuasion may seem to imply a passive role for the audience, such that the audience is assumed unable to resist the effects of narrative communication. Yet, audiences are very capable of rejecting narratives, most notably when the persuasive intent becomes salient and individuals react to the perception of being manipulated (Moyer-Guse & Nabi, 2010). Likewise, a review synthesis of narrative persuasion studies in a health context found mixed results involving other individual moderating factors (Winterbottom et al., 2008). Rather than imply a passive versus active dichotomy, where passive audiences are impacted by narrative and active audiences are not, narrative persuasion is more influenced by what type of active processing an audience chooses to use. The rejection of a narrative due to realization of its persuasive intent is an active cognitive process where an audience engages with the narrative as a communicative message. In contrast, absorption within the events of the narrative world represents a different form of active cognition that often demands enough cognitive resources to restrict other forms of thinking (Gerrig, 1993). Successful narrative persuasion therefore depends on the realization that the audience has a choice between engaging with the narrative as a message or as a world, and attempting to construct narratives that more often achieve the latter.

The relationship between narrative persuasion and science is both one of potential benefit and one of conflict. On the one hand, narratives offer a format of communication with fundamental advantages in comprehension, personal relevance, and behavior modeling. The potential of science benefiting from the persuasive use of narrative has been explored in contexts as diverse as vaccination (Brodie et al., 2001), proenvironmental beliefs (Dahlstrom, 2010, 2012, 2015), and HIV awareness (Vaughan, Rogers, Singhal, & Swalehe, 2000), with generally positive results. Outside academic research, the Centers for Disease Control and Prevention (2015) can provide an example of the perceived impact of narrative on science perceptions as it has begun working with television producers to ensure accurate portrayal of science in sit-coms and other prime-time television programs.

On the other hand, narratives have the potential to negatively impact science. Narratives do not play by the same rules as evidence-based comprehension, influencing perceptions not through spirited debate but through a whisper of suggestion. Such influence is not easily countered. In fact, accepted narratives are trusted to the extent that individuals rarely allow evidence to contradict them; the evidence is altered to fit the narratives (Shanahan & McComas, 1999). Such impacts may lead to the acceptance of incorrect scientific information or processes or the formation of negative stereotypes about scientists (Barriga, Shapiro, & Fernandez, 2010).

Using narratives in a science-policy context introduces ethical considerations that intersect with the ethical role of science in society and the ethics of science communication in general. The next section will review these issues and define the scope of our ethical examination.

Ethics in Science Communication

Because the discussion of ethics within science communication stretches across multiple domains, it is necessary to define the ethical domain in which this chapter is focused. We will first articulate four ethical domains in science communication to define the scope of our discussion and then expand upon our domain of contribution.

The first domain focuses on the ethical conduct of communication within scientific research. This domain is the focus for many STEM graduate research ethics courses, discussing the proper identification of funding sources, the disclosure of conflicts of interests, the use of informed consent to ensure no potential harm to human subjects, and the fair treatment, analysis, and reporting of research data (Horner & Minifie, 2011; Martin, 2008).

The second domain focuses on the traditional journalistic ethics of covering science. Well-established codes of ethics (Society of Professional Journalists, 2014), as well as other scholars (D. M. Cook, Boyd, Grossmann, & Bero, 2007), discuss how journalists should "seek truth and report it" by being objective, not misrepresenting factual information, not plagiarizing others' works, and avoiding conflicts of interests. As most journalists are not trained scientists and are not familiar with technical information or jargon, the mainstream press is often criticized for poor science coverage (G. Cook, Robbins, & Pieri, 2006; McInerney, Bird, & Nucci, 2004).

The third domain focuses on the coverage of ethical controversies surrounding science policy. This domain moves ethics from guiding how a journalistic story should be reported to the subject of media coverage of active ethical controversies surrounding science-related policy. Much of the ethics-related work published in *Science Communication* represents this domain, such as the examination of the autism-vaccine controversy (Clarke, 2008), stem-cell research (Leydesdorff & Hellsten, 2005), and biotechnology policy (de Cheveigne, 2002).

Our ethical discussion exists in a fourth domain that examines the use of communication techniques relative to the ethical role of science in society. While both scientific research and journalism have long histories of ethical discussion, this fourth domain, the ethics of communicating science to a nonscientist audience, has received little consideration (Meyer & Sandoe, 2012; Pimple, 2002). The relevant literature within this domain has examined the role of communication within science policy, the role of scientists within science policy, and the role of communication techniques relative to the previous contexts.

What is the role of communication within science policy? According to the Public Understanding of Science (PUS) model, controversies about science are rooted in ignorance caused by a deficit of science literacy and the role of communication is to rectify this deficit by educating the public and reducing the controversy (Miller & Kimmel, 2001; Miller, Pardo, & Niwa, 1997). Implicit in this model is that controversy is not desirable and is something that can be remedied by moving away from irrational beliefs toward factual knowledge. This model treats the public as passive vessels needing more knowledge that scientists provide and communicators translate.

Contrasting this model is the Public Engagement in Science and Technology (PEST) model, in which controversies about science represent a necessary function of the democratic process and the role of communication is to facilitate discussion about the benefits and risks of policy informed by societal values and technical information (Dickson, 2001; Walker, Simmons, Irwin, & Wynne, 1999). In this model, communication serves a two-way function, with the public actively engaging in deliberations about the benefits and risks of controversial science-related issues. Science communication scholars have shifted support from the PUS model toward the PEST model over the past decade, noting the necessary consideration of personal values and autonomy for appropriate scientific policy making (Besley, Kramer, Yao, & Toumey, 2008; Einsiedel, 2008; Irwin & Michael, 2003; Powell & Kleinman, 2008).

What is the role of scientists within political policy? Pielke (2007) discusses the possible roles that scientists can personify when contributing information toward policy-decisions. A *pure scientist* avoids commenting on policy options and instead summarizes knowledge from his or her particular field. A *science arbiter* answers technical questions about a particular policy, but avoids prescribing what should be done. An *issue advocate* does prescribe a particular policy action, aligning with one side and limiting policy options. Finally, an *honest broker* expands policy options by commenting on existing policy and offering other options that may not have appeared yet in the political agenda. Pielke (2007) claims that all four of these roles can be ethically appropriate.

What is inappropriate, according to Pielke (2007), is when a scientist claims that scientific information compels a certain policy action. In such an instance, the scientist is trying, consciously or not, to use the credibility of science to obscure the larger value debates underlying the controversy. Pielke calls this role the *stealth issue advocate* and uses climate change as an example where society is still arguing over the legitimacy of the science when the underlying value systems are driving much of the controversy. This critique about assuming science has the power to drive policy has been echoed by others (Nelson & Vucetich, 2009; Nisbet, 2009) and suggests that while there are multiple roles a scientist can play in policy debate, scientists should not use their credibility or the objectivity of science to suppress the deeper value debates intrinsic to controversial science issues.

How do specific communication techniques align with the roles of communication and of scientists within policy? The communication literature discusses many techniques used to attract, hold, inform, and persuade audiences and offers empirical tests as to their effectiveness under varied contexts. Yet, there is much less discussion as to the ethical considerations of using these communication techniques.

One field that does reflect upon the ethics of using communication techniques is advertising. As summarized by Nebenzahl and Jaffe (1998), a communication technique's ethicality is a function of the degree to which it causes harm to consumers, such as by (a) manipulating and controlling consumers' behaviors, (b) infringing on consumers' level of privacy, or (c) violating consumers' rights to be informed.

Braybrooke (1967) argued that by repeatedly showing advertisements of a certain brand or product, advertising creates a limited set of choices for consumers and manufactures a new set of desires, potentially violating consumers' autonomy by preventing them from following their rational desire or will (Braybrooke, 1967; Crisp, 1987). Likewise, exposure to advertising is not always voluntary and may happen when consumers are not conscious of their own exposure (Nebenzahl & Jaffe, 1998). Product placement where a product is seamlessly inserted into a film may therefore result in the invasion of consumers' privacy (Nebenzahl & Jaffe, 1998). Since "listeners are entitled to know by whom they are being persuaded" (Federal Communications Commission, 1963, p. 834), product placements and press releases where sponsors are not clearly stated could also be argued to be violating consumers' rights to be informed. While these ethical considerations have been discussed in an advertising context, the relation of these concerns to science communication remains relatively unexplored.

Ethical Considerations of Science Narratives

We introduce three ethical considerations a science communicator faces when deciding to use narrative as a communication technique within a policy setting. We use the neutral term "science communicator" to represent any subject who desires to communicate about a science or technical issue, realizing that this classification includes a broad range of actors including scientists, journalists, and public information officers, and we will differentiate when necessary. Likewise, our "science communicator" represents an actor with an honest desire to communicate truthfully; we are not considering an actor with the desire to spread misinformation as these actions introduce other ethical considerations and, in our perhaps optimistic opinion, represent a minority of science communicators within the larger democratic context. We also admit that much of science communication is less concerned with policy than with satisfying audience curiosity or providing information deemed useful for individual action. However, we focus on science communication within a policy context because of the frequency with which ethical challenges arise and the seriousness of their outcomes for society. Therefore, we explore the following three ethical considerations surrounding the use of narrative within a science policy context.

WHAT IS THE UNDERLYING PURPOSE FOR USING NARRATIVE: COMPREHENSION OR PERSUASION?

This consideration requires a reflection on the appropriate role of communication within science policy. Why engage in science communication? The simple answer is to "inform," but such an answer often distracts from the underlying assumptions about what informing the public is supposed to achieve. Is the purpose of informing the public to reduce controversy about a science-related policy as assumed under PUS? Or is the purpose of informing the public to facilitate the controversy necessary to reach a policy as assumed under PEST? Should science communication promote personal autonomy to make choices or create disengaged compliance toward a preferred outcome? This ongoing partition is explored elsewhere in this volume within other contexts, particularly in the chapters by Priest and Sprain. In the current context, it manifests itself in the literature between the areas of narrative persuasion and narrative comprehension.

Narratives offer benefits of persuasion through their ability to make normative claims without needing to explicitly state or defend them. These benefits are amplified through factors such as a reduced ability to counterargue when processing narrative information and identification with characters designed to exemplify the central persuasive message. A science communicator whose assumptions underlie the PUS model may decide to design a narrative with the goal of reducing a science-related controversy and generating consensus through the persuasive benefits intrinsic to narratives. Creating a narrative for these ends could involve choosing a series of events that provide a causal explanation of the preferred side of the issue, portraying those events through the eyes of a character that either normatively agrees with the preferred side of the issue, or learns to do so throughout the narrative, and either concealing or undermining the values underlying the opposing side. Such a narrative has the potential to unobtrusively persuade an audience to be more receptive to a particular science-related policy.

Narratives also offer benefits in comprehension through their increased ease of processing and ability to make information more relevant and contextual. Much of this benefit lies in the fact that narratives mirror daily experience and are therefore easier to understand and put into a human perspective. A science communicator whose assumptions underlie the PEST model may decide to design a narrative with the goal of facilitating informed debate by increasing comprehension of the science-related factors. Creating a narrative for these ends could involve selecting causal events that explain the factors underling the science issue, portraying the events through a character neutral to the issue at hand or through multiple characters to portray multiple sides, and personifying the underlying social values that intersect with the issue. Such a narrative has the potential to engage a wider public in the debate, enhance understanding of the science, and create greater connections with existing knowledge. The area of narrative medicine can provide a relevant example of using narratives for these purposes. Formed partially in response to the critique of medical personnel treating diseases instead of people, narrative medicine claims that narratives are necessary to facilitate a two-way dialogue between patient and care provider and to increase comprehension for both (Harter & Bochner, 2009).

As evidenced by attacks on the PUS model as inadequate (Einsiedel, 2008; Irwin & Michael, 2003), it can be assumed that science communication scholars would support using narrative to facilitate discussion toward informed policy. Such support raises the question of whether it is ethical to use a communication technique that must remain hidden to be effective. The persuasive impact of narrative has been demonstrated to decrease markedly once the persuasive intent becomes known (Moyer-Guse & Nabi, 2010). Yet, the improved comprehension derived from narrative is not affected if its intent becomes salient. It therefore seems safe to assume that using narrative for increased comprehension is ethically justifiable.

However, it may be naive to assume that using narrative for comprehension in a science context is appropriate while using narratives for persuasion is inappropriate, as there may be instances where manufacturing compliance represents an ethical decision. Persuasion is often the underlying purpose behind health narratives trying to promote healthy attitudes or behaviors, either as veiled entertainment programs or as carefully selected testimonials within a larger non-narrative message (Vaughan et al., 2000; Zillmann, 2006). In such cases, the social benefits of increased vaccination or environmentally conscious behaviors may justify reduced personal autonomy.

Such a decision may depend in part on the type of science issue involved. Pielke (2007) differentiates between two types of science issues, tornado politics and abortion politics. Tornado politics represent issues with high consensus where science can justify the best course of action (because everyone agrees they want to escape the tornado and wants to know how). Abortion politics represent issues with low consensus where science can never resolve a conflict of underlying values. Persuasion toward science policy may be more justifiable under tornado politics, where there is a clearly supported outcome, rather than abortion politics, where values become more contested. Likewise, the decision may also depend on the normative expectations held by the public about the communicator. Scientists and journalists are often expected to remain objective and using narrative for persuasion may take advantage of this expectation. Public information officers, on the other hand, are expected to promote their employer and persuasive narratives may be more justifiable because they are expected.

In sum, narrative can be used in a science context to increase either persuasion or comprehension. The consideration of which underlying purpose should drive the creation of a science narrative will intersect with a consideration of the appropriate role of communication within science policy.

WHAT ARE THE APPROPRIATE LEVELS OF ACCURACY TO MAINTAIN WITHIN THE NARRATIVE?

Implicit in the previous discussion is that the science being communicated in a narrative should remain accurate. Yet, unlike evidence-based communication where each fact can be individually assessed for accuracy, narratives rely on context for their meaning, which introduces differing levels of accuracy.

It may seem the most obvious measure of accuracy may be whether the narrative describes a real, nonfictional series of events or an imagined, fictional account. While nonfiction narratives may indeed be more accurate, this level of accuracy may not always be necessary. Choosing to construct a fictional narrative may sometimes be more appropriate because hypothetical situations can be created to explain relationships that have yet to occur or to model an "average" or "extraordinary" experience that might not actually occur in the complex interactions of the real world, yet would be instructive in understanding the science. Likewise, many elements within a fictional narrative can still be accurate—such as scientific procedures, cause-and-effect relationships, or probabilities of risk—and much research has demonstrated that individuals learn facts about the world just as well from fiction as nonfiction narratives (Dahlstrom, 2010; Marsh et al., 2003). To take a concrete example, consider constructing a narrative to explain the future impact of sea level rise due to climate change. Since the future impact has yet to occur, creating a nonfiction narrative would entail focusing on tangential events, such as how a particular scientist predicted the rise in sea levels or how an individual experiences the current impact of rising sea levels. Alternatively, a fictional narrative could describe how a hypothetical individual might experience the world with sea levels at the predicted height to provide a more accurate perception of the future impact than could be provided by either of the nonfiction examples.

Beyond fiction versus nonfiction, the literature on perceived realism can offer vocabulary to describe additional levels of accuracy intrinsic to narrative. Busselle and Bilandzic (2008) define narrative elements that are accurate *relative to the real world* as high on external realism. The previous examples of potentially accurate elements within a fictional narrative show how fictional narratives may nonetheless be high on external realism. Likewise, a nonfiction narrative may actually be low in external realism, such as in a historically accurate narrative that bears little resemblance to the world of today. It is likely external realism being conceptualized when scientists and scholars speak of the importance of maintaining accuracy within science communication, namely that it accurately represents the real world. One challenge, then, of creating science narratives rests in deciding what elements of the narrative need to maintain high external realism and what elements can be relaxed toward low external realism for the larger purposes of communication. This selective external accuracy is already present within science itself, such as when a scientist uses the impossible assumption of a frictionless surface or infinite plane to more clearly focus on some other aspect of reality. Some of the narrative elements that may or may not need to be high on external realism include types of characters, characters' motivations and actions, settings, situations, events, cause-and-effect relationships, procedures, chronologies, and time frames.

Again, to offer an example, consider constructing a narrative to explain the process of converting grain to ethanol by personification of the components into characters. Describing yeast as waiting until the proper temperature to eat its lunch of sugar is a cause-and-effect relationship low on external realism, but the inputs and requirements of the procedure can retain high external realism and accurately describe the process in an understandable form.

In contrast to external realism, Busselle and Bilandzic (2008) define elements of a narrative that are accurate *relative to the rules set forth within the narrative world* as high on narrative realism. Rules are established early in every narrative about how the narrative world operates, such as how a character relates to the world around him or her, the properties of an object, or the importance of certain objectives. If these rules are later broken, processing is hindered (Albrecht & Obrien, 1993; Kaup & Foss, 2005), persuasion is decreased (Green, 2006; Green & Brock, 2000), and the narrative may even be rejected (Busselle & Bilandzic, 2008; Hall, 2003). The influence of narrative realism becomes obvious in the case of fantasy, where a dragon breathing fire is seen as more accurate than a dragon breathing water, even though both are unquestionably externally inaccurate. Narrative realism represents a second level of accuracy that is less often discussed in science communication, but the implications remain. A narrative that maintains accuracy through appropriate external realism may nonetheless be rejected because it is not accurate as perceived through narrative realism. For instance, constructing a persuasive narrative to promote the acceptance of genetically modified food by showing a character benefit from their use may maintain appropriate external realism, yet be perceived as narratively inaccurate if the character does not behave as earlier descriptions would suggest or if the narrative does not complete a story arc and feels incomplete.

Yet another level of accuracy within narrative communication lies in a narrative's representativeness. Narratives intrinsically lead to the abstracting of specific examples to general trends. Unlike the generalizable content of evidence-based communication, narratives provide a single or small number of exemplars relative to an issue. The representativeness of these exemplars will therefore determine the accuracy with which an audience can generalize to other contexts. Exemplification theory clearly shows the power of narrative to impact perceptions of representativeness. Even when base rate information is present claiming a particular risk is low, the presence of exemplars skews perceptions toward the typicality of the specific exemplar used (Gibson & Zillmann, 1994; Zillmann, 2006). This third level of accuracy raises the possibility of creating a narrative that maintains accuracy through appropriate external and narrative realism, but fails to accurately depict a series of events that is representative of the larger science issue. For instance, sharing a narrative of the experience of an individual who decided not to be vaccinated and then developed polio may maintain the desired previous levels of accuracy, but also represents a worst-case scenario that is not generalizable to what is likely to occur in such a situation and is therefore representationally inaccurate. Of course, selecting a nonrepresentative narrative could be beneficial for a science communicator attempting to use narrative to persuade an audience toward a predetermined end (here, vaccination).

While accuracy remains crucial for appropriate science communication, the maxim of maintaining accuracy becomes more complex when considering narrative communication. While maintaining accuracy through narrative realism is likely necessary for any effective narrative, the choice in constructing a fiction versus nonfiction narrative, what elements of the narrative should exhibit high external realism, and whether to select a representative example will intersect with the science issue at hand, the nature of what is to be communicated, and the underlying purpose for using narrative in a science context.

SHOULD NARRATIVE BE USED AT ALL?

Whereas some communication techniques are unavoidable, narratives do represent a communicative choice—a choice between using evidence-based arguments through paradigmatic processing or a mediated experience through narrative processing. What are the ethical considerations of using one cognitive pathway over the other? It now becomes necessary to examine specific actors within our broad "science communicator" label. Journalists, public information officers, and other roles that specialize in communication are often expected to use narrative. In fact, the *Field Guide for Science Writers* specifically cites narrative as one of the effective techniques for covering science (Blum, Knudson, & Henig, 2006).

Scientists, however, hold a much different role in society and their use of narrative raises its own considerations. Do the perceptions of scientists so closely align with evidence-based communication of the paradigmatic pathway that scientists will be perceived as violating normative expectations if they dabble in the narrative pathway? Could such normative violations cause particular scientists to lose their credibility in the public sphere, or worse, cause the science itself to lose credibility? Bruner (1986) expands on this idea, saying that a scientist caught using the presupposition or subjectification common in narrative would become the "butt of jokes" (p. 28) while a novelist could not maintain suspense or reader involvement without them (Bruner, 1986). While successful scientist-popularizers, such as Carl Sagan, are often championed as the cure for poor science literacy (Nisbet & Scheufele, 2009), a 2006 survey of scientists' views of engaging with nonscientists reveal that scientists who communicate with nonscientists are often viewed by their peers as "fluffy" or "not good enough" for an academic career (The Royal Society, 2006). There is yet no research into what normative expectations the public holds regarding the communication of scientists to a nonscientist audience, so these questions cannot be addressed here.

Nonetheless, other communicators within the debate will likely use narratives, and a choice to avoid using narratives completely due to normative expectations may represent a capitulation to those who do. This dilemma may be becoming more salient as exemplified in a 2008 speech where National Public Radio science correspondent Robert Krulwich criticized scientists for not using their own narratives to counter the "beautiful" narratives of creationism and told young scientists that they should tell stories as a way to fight back (Krulwich, 2008). Regardless of any existing normative expectations aligning scientists to the paradigmatic mode of communication, it may, in fact, be unethical for scientists *not* to use narrative and to surrender the benefits of a communication technique to the non-expert side of an issue.

The question of whether or not to use narrative at all in a science-policy context is essentially a philosophical question. While all of the considerations thus far are relevant when evaluating the question on a situational level, it may be illustrative to ground the considerations in contrasting ethical theories. Utilitarianism is one such theory that focuses on the aspect of harm and states that ethical actions are those that produce the greatest good for the greatest number of people (Rachels & Rachels, 2007). Following utilitarianism, the use of narrative is ethically justified anytime the effects benefit more people than it harms. In contrast, Kant's categorical imperative focuses on autonomy and states that ethical actions are those that respect an individual's rationality and do not treat others as a means to an end (Kant, 1785/1981). Following this aspect of Kant's categorical imperative, the use of narrative is ethically justified anytime it does not attempt to restrict an individual's autonomy to make decisions.

To illustrate the contrasts between these two ethical theories, let us examine two possible questions about the use of narrative: (a) are narratives always manipulative and (b) when is manipulation appropriate? The first question would be of utmost importance under Kant's categorical imperative as narratives could be ethically justified only if the answer was no. Under utilitarianism, the first question is of little consequence yet the second question demands careful consideration of the outcome of the manipulation. Returning to Kant's categorical imperative, the second question has a simple answer: never.

It is not realistic to espouse a particular ethical theory on which to ground a proposed ethical approach to using narrative. Not only is it suggested that reliance on a single ethical theory is not how individuals actually make decisions (Daniels, 1979), but also it oversimplifies the complexities that exist when communicating science in a policy context in which conflicting viewpoints are present. It is more important to emphasize that specific instances of communication are situational, and the decision whether or not to use a narrative should be one of the considerations.

Conclusion

Using narratives within a science policy context offers a range of benefits related to comprehension and interest, yet also presents a set of ethical considerations not present in more traditional science communication formats. The goal of this chapter is not to argue for or against narrative use in this context, nor to suggest what narratives would be most impactful. Instead, our goal is to articulate what is at stake when a communicator pauses to consider whether or how to inject a narrative message into a complex science policy situation. As some areas of science become more politicized, the desire to influence public opinion is likely to increase. We hope that science communicators can use the considerations outlined in this chapter to make thoughtful communication choices that align with their own ethical standards.

References

AAAS. (2016). Communicating science workshops. Retrieved from http://www.aaas.org/pes/communicating-science-workshops

Albrecht, J. E., & Obrien, E. J. (1993). Updating a mental model—Maintaining both local and global coherence. *Journal of Experimental Psychology—Learning Memory and Cognition, 19*(5), 1061–1070.

Appel, M., & Richter, T. (2007). Persuasive effects of fictional narratives increase over time. *Media Psychology, 10*(1), 113–134.

Baker, J. P. (2008). Mercury, vaccines, and autism—One controversy, three histories. *American Journal of Public Health, 98*(2), 244–253.

Barriga, C. A., Shapiro, M. A., & Fernandez, M. L. (2010). Science information in fictional movies: Effects of context and gender. *Science Communication, 32*(1), 3–24.

Basken, P. (2009). Often distant from policy making, scientists try to find a public voice. *Chronicle of Higher Education, 55*(38).

Besley, J. C., Kramer, V. L., Yao, Q., & Toumey, C. (2008). Interpersonal discussion following citizen engagement about nanotechnology: What, if anything, do they say? *Science Communication, 30*(2), 209–235.

Blum, D., Knudson, M., & Henig, R. M. (2006). *A field guide for science writers* (2nd ed.). New York, NY: Oxford University Press.

Braddock, K., & Dillard, J. P. (2016). Meta-analytic evidence for the persuasive effects of narrative on beliefs, attitudes, intentions, and behaviors. *Communication Monographs, 83*(4), 446–467.

Braybrooke, D. (1967). Skepticism of wants, and certain subversive effects of corporations on American values. In H. Sidney (Ed.), *Human values and economic policy.* New York: New York University Press.

Brodie, M., Foehr, U., Rideout, V., Baer, N., Miller, C., Flournoy, R., & Altman, D. (2001). Communicating health information through the entertainment media. *Health Affairs, 20*(1), 192–199.

Bruner, J. (1986). *Actual minds, possible worlds.* Cambridge, MA: Harvard University Press.

Bruner, J. (1991). The narrative construction of reality. *Critical Inquiry, 18*(1), 1–21.

Busselle, R. W., & Bilandzic, H. (2008). Fictionality and perceived realism in experiencing stories: A model of narrative comprehension and engagement. *Communication Theory, 18*(2), 255–280.

Centers for Disease Control and Prevention. (2015). "Entertainment education." Retrieved from http://www.cdc.gov/healthcommunication/ToolsTemplates/EntertainmentEd/

Clarke, C. E. (2008). A question of balance: The autism-vaccine controversy in the British and American elite press. *Science Communication, 30*(1), 77–107.

Cook, D. M., Boyd, E. A., Grossmann, C., & Bero, L. A. (2007). Reporting science and conflicts of interest in the lay press. *PLOS ONE, 2*(12). doi:10.1371/journal.pone.0001266

Cook, G., Robbins, P. T., & Pieri, E. (2006). "Words of mass destruction": British newspaper coverage of the genetically modified food debate, expert and non-expert reactions. *Public Understanding of Science, 15*(1), 5–29.

Crisp, R. (1987). Persuasive advertising, autonomy, and the creation of desire. *Journal of Business Ethics, 6*(5), 413–418.

Curtis, R. (1994). Narrative form and normative force: Baconian storytelling in popular science. *Social Studies of Science, 24*(3), 419–461.

Dahlstrom, M. F. (2010). The role of causality in information acceptance in narratives: An example from science communication. *Communication Research, 37*(6), 857–875.

Dahlstrom, M. F. (2012). The persuasive influence of narrative causality: Psychological mechanism, strength in overcoming resistance, and persistence over time. *Media Psychology, 15*(3), 303–326.

Dahlstrom, M. F. (2014). Using narratives and storytelling to communicate science with non-expert audiences. *Proceedings of the National Academy of Sciences of the United States of America, 111,* 13614–13620.

Dahlstrom, M. F. (2015). The moderating influence of narrative causality as an untapped pool of variance for narrative persuasion. *Communication Research, 42*(6), 779–795.

Dal Cin, S., Zanna, M. P., & Fong, G. T. (2004). Narrative persuasion and overcoming resistance. In E. Knowles & J. Linn (Eds.), *Resistance and persuasion* (pp. 175–191). Mawah, NJ: Erlbaum.

Daniels, N. (1979). Wide reflective equilibrium and theory acceptance in ethics. *Journal of Philosophy, 76*(5), 256–282.

de Cheveigne, S. (2002). Biotechnology policy—Can France move from centralized decision making to citizens' governance? *Science Communication, 24*(2), 162–172.

de Wit, J. B. F., Das, E., & Vet, R. (2008). What works best: Objective statistics or a personal testimonial? An assessment of the persuasive effects of different types of message evidence on risk perception. *Health Psychology, 27*(1), 110–115.

Dickson, D. (2001). Weaving a social web—The internet promises to revolutionize public engagement with science and technology. *Nature, 414*(6864), 587–587.

Durfee, J. L. (2006). "Social change" and "status quo" framing effects on risk perception: An exploratory experiment. *Science Communication, 27*(4), 459–495.

Einsiedel, E. (2008). Public engagement and dialogue: A research review. In M. Bucchi & B. Smart (Eds.), *Handbook of public communication on science and technology* (pp. 173–184). London, UK: Routledge.

Federal Communications Commission. (1963). FCC report (Vol. 34, pp. 834–835). Washington, DC: FCC. Retrieved from https://digital.library.unt.edu/ark:/67531/metadc177313/

Fisher, W. R. (1984). Narration as a human-communication paradigm: The case of public moral argument. *Communication Monographs, 51*(1), 1–22. Retrieved from http://www.tandf.co.uk/journals/rcmm/

Forrest, B. (2001). The wedge at work: How intelligent design creationism is wedging its way into the cultural and academic mainstream. In R. T. Pennock (Ed.), *Intelligent design creationism and its critics* (pp. 5–53). Cambridge, MA: MIT Press.

Gerrig, R. J. (1993). *Experiencing narrative worlds: On the psychological activities of reading.* Boulder, CO: Westview Press.

Gibson, R., & Zillmann, D. (1994). Exaggerated versus representative exemplification in news reports—Perceptions of issues and personal consequences. *Communication Research, 21*(5), 603–624.

Golding, D., Krimsky, S., & Plough, A. (1992). Evaluating risk communication—Narrative vs technical presentations of information about radon. *Risk Analysis, 12*(1), 27–35.

Graesser, A. C., Olde, B., & Klettke, B. (2002). How does the mind construct and represent stories? In M. C. Green, J. J. Strange, & T. C. Brock (Eds.), *Narrative impact: Social and cognitive foundations* (pp. 229–262). Mahwah, NJ: Lawrence Erlbaum.

Green, M. C. (2004). Transportation into narrative worlds: The role of prior knowledge and perceived realism. *Discourse Processes, 38*(2), 247–266.

Green, M. C. (2006). Narratives and cancer communication. *Journal of Communication, 56,* S163–S183.

Green, M. C., & Brock, T. C. (2000). The role of transportation in the persuasiveness of public narratives. *Journal of Personality and Social Psychology, 79*(5), 701–721.

Hall, A. E. (2003). Reading realism: Audiences' evaluations of the reality of media texts. *Journal of Communication, 53*(4), 624–641.

Harris, G., & O'Conner, A. (2005, June 25). On autism's cause, it's parents vs. research. *The New York Times.* Retrieved from http://www.nytimes.com/2005/06/25/science/25autism.html

Harter, L. M., & Bochner, A. P. (2009). Healing through stories: A special issue on narrative medicine. *Journal of Applied Communication Research, 37*(2), 113–117.

Hertog, J. K., & McLeod, D. M. (2001). A multiperspectival approach to framing analysis: A field guide. In S. Reese, O. Gandy, & A. Grant (Eds.), *Framing public life: Perspectives on media and our understanding on the social world* (pp. 139–162). Mahwah, NJ: Lawrence Erlbaum.

Hinyard, L. J., & Kreuter, M. W. (2007). Using narrative communication as a tool for health behavior change: A conceptual, theoretical, and empirical overview. *Health Education & Behavior, 34*(5), 777–792.

Horner, J., & Minifie, F. D. (2011). Research ethics II: Mentoring, collaboration, peer review, and data management and ownership. *Journal of Speech Language and Hearing Research, 54*(1), S330–S345.

Irwin, A., & Michael, M. (2003). *Science, social theory and public knowledge.* Philadelphia, PA: Open University Press.

Kant, I. (1981). Grounding for the metaphysics of morals. J. W. Ellington (Trans.). Indianapolis, IN: Hackett. (Original work published 1785)

Kaup, B., & Foss, D. (2005). Detecting and processing inconsistencies in narrative comprehension. In D. Rosen (Ed.), *Trends in experimental psychology research* (pp. 67–90). Hauppauge, NY: Nova Science.

Kreiswirth, M. (1992). Trusting the tale: The narrativist turn in the human sciences. *New Literary History, 23*(3), 629–657.

Krulwich, R. (2008, July 29). Tell me a story [Audio podcast]. Retrieved from http://www.radiolab.org/blogs/radiolab-blog/2008/jul/29/tell-me-a-story/

Leydesdorff, L., & Hellsten, N. (2005). Metaphors and diaphors in science communication. *Science Communication, 27*(1), 64–99.

Liu, H., & Priest, S. H. (2009). Understanding public support for stem cell research: Media communication, interpersonal communication and trust in key actors. *Public Understanding of Science, 18*(6), 704–718.

Longacre, R. (1983). *The grammar of discourse.* New York, NY: Plenum Press.

Marsh, E. J., Meade, M. L., & Roediger, H. L. (2003). Learning facts from fiction. *Journal of Memory and Language, 49*(4), 519–536.

Martin, D. W. (2008). *Doing psychology experiments* (7th ed.). Singapore: Wadsworth.

McInerney, C., Bird, N., & Nucci, M. (2004). The flow of scientific knowledge from lab to the lay public—The case of genetically modified food. *Science Communication, 26*(1), 44–74.

Meyer, G., & Sandoe, P. (2012). Going public: Good scientific conduct. *Science and Engineering Ethics, 18*(2), 173–197.

Miller, J. D., & Kimmel, L. (2001). *Biomedical communications: Purposes, audiences, and strategies.* New York, NY: John Wiley.

Miller, J. D., Pardo, R., & Niwa, F. (1997). *Public perceptions of science and technology: A comparative study of the European Union, the United States, Japan, and Canada.* Chicago, IL: Chicago Academy of Sciences.

Mink, L. O. (1978). Narrative form as a cognitive instrument. In R. H. Canary (Ed.), *The writing of history* (pp. 129–149). Madison: University of Wisconsin Press.

Mooney, C. (2005). *The Republican war on science.* New York, NY: Basic Books.

Moyer-Guse, E., & Nabi, R. L. (2010). Explaining the effects of narrative in an entertainment television program: Overcoming resistance to persuasion. *Human Communication Research, 36*(1), 26–52.

Nebenzahl, I. D., & Jaffe, E. D. (1998). Ethical dimensions of advertising executions. *Journal of Business Ethics, 17*(7), 805–815.

Nelson, M. P., & Vucetich, J. A. (2009). On advocacy by environmental scientists: What, whether, why, and how. *Conservation Biology, 23*(5), 1090–1101. doi:10.1111/j.1523–1739.2009.01250.x

Nisbet, M. C. (2009). The ethics of framing science. In B. Nerlich, B. Larson, & R. Elliott (Eds.), *Communicating biological sciences: Ethical and metaphorical dimensions* (pp. 51–73). London, UK: Ashgate.

Nisbet, M. C., & Scheufele, D. A. (2009). What's next for science communication? Promising directions and lingering distractions *American Journal of Botany, 96*(10), 1767–1778.

Oatley, K. (1999). Why fiction may be twice as true as fact: Fiction as cognitive and emotional simulation. *Review of General Psychology, 3*(2), 101–117.

Pielke, R. A. (2007). *The honest broker: Making sense of science in policy and politics.* Cambridge, UK: Cambridge University Press.

Pimple, K. (2002). Six domains of research ethics. *Science and Engineering Ethics, 8*(2), 191–205.

Powell, M., & Kleinman, D. L. (2008). Building citizen capacities for participation in nanotechnology decision-making: The democratic virtues of the consensus conference model. *Public Understanding of Science, 17*(3), 329–348.

Rachels, J., & Rachels, S. (2007). *The elements of moral philosophy.* New York, NY: McGraw Hill.

Read, S. J., & Miller, L. C. (1995). Stories are fundamental to meaning and memory: For social creatures, could it be otherwise? In R. Wyer (Ed.), *Knowledge and memory: The real story* (pp. 139–152). Hillsdale, NJ: Lawrence Erlbaum.

The Royal Society. (2006). Factors affecting science communication. Retrieved from http://royalsociety.org/Content.aspx?id=5232

Schank, R. C., & Abelson, R. (1995). Knowledge and memory: The real story. In R. C. Schank & R. Abelson (Eds.), *Knowledge and memory: The real story* (pp. 1–86). Hillsdale, NJ: Lawrence Erlbaum.

Shanahan, J., & McComas, K. (1999). *Nature stories: Depictions of the environment and their effects.* Cresskill, NJ: Hampton Press.

Slater, M. D., & Rouner, D. (2002). Entertainment-education and elaboration likelihood: Understanding the processing of narrative persuasion. *Communication Theory, 12*(2), 173–191.

Slater, M. D., Rouner, D., & Long, M. (2006). Television dramas and support for controversial public policies: Effects and mechanisms. *Journal of Communication, 56*(2), 235–252.

Society of Professional Journalists. (2014). SPJ code of ethics. Retrieved from http://www.spj.org/ethicscode.asp

Vaughan, P. W., Rogers, E. M., Singhal, A., & Swalehe, R. M. (2000). Entertainment-education and HIV/AIDS prevention: A field experiment in Tanzania. *Journal of Health Communication, 5*, 81–100.

Volti, R. (2009). *Society and technological change* (6th ed.). New York, NY: Worth.

Walker, G., Simmons, P., Irwin, A., & Wynne, B. (1999). Risk communication, public participation and the Seveso II directive. *Journal of Hazardous Materials, 65*(1–2), 179–190.

Winterbottom, A., Bekker, H. L., Conner, M., & Mooney, A. (2008). Does narrative information bias individual's decision making? A systematic review. *Social Science & Medicine, 67*(12), 2079–2088.

Zillmann, D. (2006). Exemplification effects in the promotion of safety and health. *Journal of Communication, 56*, S221–S237.

Zimmer, C. (2011). *Evolution*. New York, NY: Harper Collins.

6

Science Communication as Communication about Persons

BRENT RANALLI

Science communication might seem one of the most *impersonal* genres of communication. Yet science communication is also, always, communication about persons. Even the most dry and abstract research paper projects the author's self-image of competence, credibility, and authority. And in other genres, such as science journalism, the person of the scientist often takes center stage.

This chapter explores the ethical dimension of communication about persons in science communication. Many persons are involved in the production of scientific knowledge; our focus will be primarily (though not exclusively) on the person of the scientist. Science communication is understood to extend from the most seemingly impersonal types of writing by and about science and scientists (e.g., the research article and textbook) to the most obviously and intensely personal (e.g., biography and autobiography).

The chapter draws insights from multiple disciplines. When it comes to articulating general ethical principles, we look to those whose business it is to think and write about persons: historians and literary theorists who practice and comment upon the discipline of biography. We look to sociology, however, for an understanding of the embedding of personal content in *all* communication, including the most formal science communication, and to philosophy and social studies of science for an account of what is at stake: why communication about persons matters in science. Throughout, I bear in mind that as first-person communication by scientists reflects on the character of the scientist, so third-person communication *about* scientists reflects equally on the character of the science writer. I conclude by offering some general observations about this parallelism: the applicability of what are conventionally considered "scientific" norms to all scholarship, and the possibility that science can learn from the ethics of humanistic disciplines.

All Communication (including Science Communication) Is Communication about Persons

In the early days of science studies, P. B. Medawar (1963) asked, "Is the Scientific Paper a Fraud?" and concluded that, in a generic sense, it is. The conventional style of scientific writing obscures the existence of an author, a human being driven by passions and interests.

That scientists write in a "barbarously impersonal style" (in the words of science scholar John Ziman; 2002, p. 40) shuts us off from access to the human side of science. But even as it hides, it also reveals. By writing in an impersonal manner, the scientist signals allegiance and a personal commitment to shared ideals of objectivity and disinterestedness (Gross, Harmon, & Reidy, 2002, p. 215). Further, by writing in the communally accepted manner, the scientist presents him- or herself as competent and self-controlled, fit to be a member of the community. By making new truth claims the scientist presents him- or herself as industrious and brilliant, or at least technically competent. Thus the formal scientific literature is brimming with subtexts about the authors who contribute to it.

Sometimes the efforts of social scientists to "strip away the mask" of objectivity and emotional neutrality are perceived as attacks on science, as if they were exposing hypocrisy among scientists (consider Medawar's provocative use of the word "fraud"). In many or most cases they need not be interpreted this way. For a passionate person to put on a professional mask of disinterestedness is not fraud, it is just the way things are done (see Shapin, 2010, chs. 2–3; Ziman, 2002, ch. 3).

In truth, the scientist could hardly fail to put on some sort of mask. The sociologist Erving Goffman (1959) observed that every social interaction is a presentation of the self to the other, a presentation that expresses an intention or wish to be seen or viewed or judged in one way rather than another. If, for the sake of argument, a scientist were to attempt to strip off the mask of impersonal disinterestedness and to present his or her research in the form of an intimate personal confession, striving for the rawest Rousseavian authenticity in Proustian detail, he or she *still* would be presenting a mask, just a mask of a different sort.[1]

---

1. Goffman (1959, p. 250) intimated that there are "gentlemanly" ways of presenting oneself for viewing simply as one is, with no pretense. This does not seem credible. If one is conscious

Arguably, an intimately personal manner of self-presentation would impede effective communication among scientists. If some sort of mask is inevitable, the mask of impersonality that is customary in science might be thought to serve a useful purpose.

In a third-person account, the person of the scientist may be present explicitly or only implicitly when his or her work is discussed (in much the same manner as the person of the scientist is implicitly present in his own formal research paper). In the most abstract textbook situation, where what is being presented is not the work of identifiable individuals but the legacy of the scientific community at large, the scientific person may vanish. But the qualities and characteristics of the scientific community as a *collective of persons* are implicitly present. And this is not trivial. In many kinds of third-person writings about science, generalizations are made or evidence provided about scientific communities' collective brilliance or civility or otherworldliness or avarice or open-mindedness or closed-handedness. Such collective qualities matter as much as individual qualities, as we will see in the next section.

Also, we should not forget that even in third-person accounts of science and scientists, it is still the first person who is speaking (to paraphrase Henry D. Thoreau, 1971, p. 3). When the journalist or sociologist or biographer writes about science, she is presenting not only her subject to the reader, but also herself. The journalist, sociologist, and biographer, too, are liable to be seen and judged whether the "I" is included or omitted.

Why Communication about Scientific Persons Matters

SELF-PRESENTATION: AUTHORITY AND CREDIBILITY

Scientists, being in the business of discovering and certifying truth, are constantly engaged in the task of establishing credibility and authority. Scientific claims are strongly backstopped by replicability of results, but no scientist is ever capable of personally verifying more than a fraction of the information he or she takes in from colleagues, and a layperson is typically unequipped

---

of being viewed, the logic of Goffman's theory would imply that even supposedly "gentlemanly" behavior amounts to a mere show. Arguably, if one were completely unconscious of being viewed and judged, the self on display might be considered more "authentic"—but only in the sense of not being deliberately adopted to manipulate this particular audience. It would still be a habitual persona of one sort or another. The "mask" metaphor, ultimately, is flawed. There is no masking and unmasking, merely the presenting of various countenances. Note that Goffman was primarily concerned with self-presentation in face-to-face interaction, but his insights apply equally well to self-presentation in prose.

to personally verify any cutting edge science at all. We all take a vast amount of knowledge, including scientific knowledge, on trust (Shapin, 2010, ch. 2).

What factors make a truth-claim credible in the eyes of the beholder? One factor that is within the speaker's control is the medium and packaging of the information in that medium: the traditional domain of rhetoric. Another that is entirely out of the speaker's control is the "fit" of the new information with the second party's existing stock of knowledge and belief (Haack, 1993). Still another, only partly in the control of the speaker, is the speaker's own perceived authority and credibility. Authority and credibility are necessarily audience-specific (e.g., Kahan & Braman, 2006), but there are some personal characteristics that appear to be widely shared in our culture as markers of a "good scientist."

The best-known formulation of these markers is sociologist Robert K. Merton's doctrine of "scientific norms" (Merton & Storer, 1973; Ziman, 2002). Merton's norms include *universalism* (participation open to all, contributions to be judged on technical merits rather than the race, religion, nationality, etc., of the person who contributed them), *disinterestedness* (judgment not to swayed by egoistic or material interests), *communalism* (open sharing of methods and results), and *skepticism* (critical scrutiny to be applied widely, no sacred cows).[2] These four norms capture many of the most important expectations we have of scientists' behavior and comportment.

In a recent study of the rhetoric of the climate science culture wars, I found that both climate activists and climate skeptics share fairly consistent expectations about what kind of persons they expect and want scientists to be. They want their scientists to be humble (self-effacing, open-minded, willing to critically revise their own opinions), and at the same time skeptical (hard-nosed, independent-minded, ready to challenge the work of others) (Ranalli, 2012a). These desired personal qualities can be seen as correlating roughly with the Mertonian norms of disinterestedness and skepticism.

To take an illustrative example, here is an excerpt from a posting on Watts Up With That, the blog of climate skeptic Anthony Watts (2011):

> [Scientist A], another affable Canadian from Toronto, with an office covered in posters to remind him of his roots, has not even a hint of the arrogance and advance certainty that we've seen from people like [Scientist B]. [Scientist A is] much more like [Scientist C] in his demeanor and approach. In

2. Merton, unfortunately, did not provide precise definitions of the terminology he introduced. Subsequent scholars have adapted them in different ways. The definitions I have provided are intended to be representative of the way the terms are commonly used in the literature and consistent with Merton's own usage.

fact, the entire team seems dedicated to providing an open source, fully transparent, and replicable method no matter whether their new metric shows a
trend of warming, cooling, or no trend at all, which is how it should be. . . .
[Scientist D] hasn't been very outspoken, which is why few people have heard
of him. I met with him and I can say that [Scientist E], [Scientist F], [Scien-
tist G], or [Scientist B] he isn't. What struck me most about [Scientist D],
besides his quiet demeanor, was the fact that it was he who came up with a
method to deal with one of the greatest problems in the surface temperature
record that skeptics have been discussing. His method, which I've been given
in confidence and agreed not to discuss, gave me one of those "Gee whiz, why
didn't I think of that?" moments. So, the fact that he was willing to look at the
problem fresh, and come up with a solution that speaks to skeptical concerns,
gives me greater confidence that he isn't just another [Scientist F] or [Scien-
tist G] re-run.

Here judgments are being made about the character of scientists based on
visible and tactile evidence (wall furnishings that suggest humility), personal
demeanor, actions taken, and words spoken. And in the shorthand of the
skeptical community, the names of certain scientists are taken to stand for
particular character traits such as stubbornness or arrogance. None of this is
to say that character is the *only* yardstick, or even the primary one, by which
skeptics—or their intellectual opponents—judge the quality of contributions to science. But the evidence from the blogosphere indicates that character is *one* heuristic yardstick that laypersons do use to judge the quality of
scientists' contributions, and to explain and justify their judgments.[3]

To Merton's well-established doctrine of norms, this research adds two
insights. First, it is possible and sometimes useful to make character (qualities of persons) rather than norms (rules and standards of behavior) the focus
of inquiry. Second, the virtues exhibited by and attributed to scientists serve
not only to enhance (or detract from) the credibility of scientists as individuals. They also enhance (or detract from) the credibility of entire scientific
communities. That is, if a community of scientists reaches consensus on a
particular point, the public is likely to judge the reliability of that consensus
at least in part on the extent to which the consensus is "hard-won" rather
than attributable to groupthink or bias or interest—and an important heuristic basis for evaluating the quality of a consensus is the character of individual members of the community. If individual scientists are (or appear to
be) timid or conformist rather than independent-minded, for example, the

---

3. And not only laypersons: Scientists, too, judge fellow scientists on the basis of personal
qualities (see Ranalli, 2012a, p. 188 and fn. 5).

consequence may be that they will fail (or will appear to fail) to challenge each others' views and seriously explore alternative hypotheses, making the consensus they achieve suspect (Ranalli, 2012a).

Goffman (1959) writes about social self-presentation as not only the concern of individuals, but of teams of individuals, who "band together and directly manipulate the impression that they give" as a "performing team" (p. 251). This could be a description of the scientific community seeking to uphold its collective credibility and authority. And this effort too need not be understood pejoratively. In making a show of being independent-minded rather than conformist, scientists behave independent-mindedly and reap the benefits of independent-mindedness: quality control of published findings, generation of new hypotheses, etc. Put another way, putting on a show of being independent-minded is *constitutive* of independent-mindedness.

SELF-DEFINITION: MORAL SELF-AUTHORSHIP

Though the personal dimension of first-person communication is partly a matter of image management, the contributions of virtue ethicists Charles Taylor and Alasdair MacIntyre show that it is also something more essential: moral self-authorship.

Whereas deontological ethics seeks to answer the question, "What is it right to do?" virtue ethics asks, "What is it good to be?" (Parker, 2004, p. 53). Taylor and those who work in his tradition emphasize that "as moral beings . . . we grasp our lives as narratives," where narratives are understood to range from "the mostly implicit stories we have of where we are 'at' in our lives" to full-fledged formal autobiography (Parker, 2004, pp. 57–58). Further, the self is "something which can exist only in a space of moral issues" (Eakin, 2004, p. 14). In this view, all communication about the self—that is, all communication—"cannot but reflect the self's orientation, or pattern of orientation, in moral space" (Parker, 2004, p. 61).

If Taylor teaches that we are continually confronted with the question "What is it good to be?" MacIntyre (1978, 1984) explains what form the answer must take. He argues that what it means to *be good* is role-specific. To be a good chess-player, or parent, or warrior, or scientist, requires cognitive work—understanding and internalizing the written and unwritten rules and approved goals—and moral work—striving excellently to achieve those goals within the given constraints.

What this tradition offers to us is the insight that the self-definition done by scientists in their formal and informal writings (and a thousand other ways) is not only self-presentation done to create an impression in others of competence and credibility and authority. It is normally internally motivated as well: the product of an individual striving, in the best way he or she knows how, to fulfill a socially defined role that he or she has voluntarily adopted as a life project (Söderqvist, 1997).

FOR THE CITIZEN, A MORE ACCURATE UNDERSTANDING OF THE SCIENTIFIC ENTERPRISE

What is at stake for the lay public in the personal dimension of science communication? Perhaps most importantly, candid writing about the lives of people who engage in science *humanizes* and *demythologizes* science. It helps people orient themselves in their universe by providing a more accurate and clear-eyed view of this important institution than they might otherwise come to if all they had was a schoolbook understanding of the "scientific method" and an appreciation of the amazing technical marvels and immense destructive power that flow from scientific laboratories. Humanizing science can relieve the semi-divine aura that all too easily can come to surround an institution with such immense cultural authority. A clear picture of the human dimension of science would be particularly valuable for the student who might contemplate a career in science.

FOR THE SCIENCE SCHOLAR, AN ADDITIONAL DIMENSION OF INSIGHT

For those who wish to know how science works—and this category includes both the professional ranks of humanists and social scientists and the interested lay public—an understanding of scientists as people is indispensable.

That indispensability hasn't always been appreciated. For much of the twentieth century, serious historians of science eschewed biography as an inferior form of scholarship. This prejudice was born out of antipathy for the moralistic "great man" narrative that was typical of scientific biography in the nineteenth century (see Söderqvist, 2007, p. 244ff for a historical overview of the genre). For much of the twentieth century, the essence of science was sought in methodology, or conceptual commitments, or social norms. As each of these approaches has failed to entirely satisfy, the science studies community has largely fallen back on a pragmatic definition of science as simply "what scientists do." Greater interest has followed in scientists as people, indeed scientists as (in the inelegant title of one excellent book on the subject) "people with bodies, situated in time, space, culture, and society, and struggling for credibility and authority" (Shapin, 2010).

Correspondingly, biography—the study of persons—has found renewed scholarly respectability.[4] And the new willingness to view science as one component in the larger context of scientists' lives shows great promise. The Danish scholar Thomas Söderqvist found the stormy personal life of twentieth-century immunologist Niels Jerne such a rich mine for interpreting and explaining the origins of Jerne's scientific theories that he saw fit to use the motto "Science as Autobiography" as the title of Jerne's biography. Söderqvist (2003) explains that Jerne's creative theorizing about the nature of the immune system, later validated by empirical research, "was, to use Nietzsche's words, 'the confession of its originator, and a species of involuntary and unconscious auto-biography'" (p. xxv).

Söderqvist (1996) offers physicist Percy Bridgman (p. 60) and zoologist Ilya Metchnikoff (p. 70) as additional examples of scientists whose creative professional achievements can be best appreciated in the context of their personal existential struggles. It may not be in every case that a scientist's contributions are so transparently autobiographical. But personal stories will undoubtedly provide insight in many cases.

FOR DISPUTANTS, HUMANIZING THE "OTHER"

Reading about persons offers a reflective opportunity to step into others' shoes. As literary scholar Larry Lockridge (1999) writes, "readers of biographies are tacitly engaged in a continuous hypothetical testing of whether a substitution could be made of their own identity for that of the biographical subject. . . . Readers of biographies in effect ask themselves, . . . 'Could I conceivably do that? Could I be that?'" (p. 136). Reading about others provides an opportunity to expand our moral universe through empathy and imagination (Eakin, 2004, p. 14). It allows us to appreciate our common humanity in spite of superficial differences, and it also may open our eyes to ways of being that are quite different from those we expect or consider the norm.

Consider an example from the climate science wars: One of the memes that is popular in climate-skeptical circles is a suspicion, or conviction, that climate scientists are "only in it for the money." This, it seems to me, reflects naiveté not only about how science funding works but also about the character of scientists. Greed is far from the characteristic vice of scientists.

---

4. While well-written biography today eschews the "great man" narrative in favor of nuance and context (including the roles played by colleagues, assistants, family members, etc. in a scientist's professional life, for example), simplistic celebration of famous men and women remains a common trope in textbooks and remains a temptation for science journalists.

Scientists may be subject to pride and perhaps envy as well, each wishing to earn honors and set a mark on history. But on the whole, greed would appear to rank fairly low among scientists' motivations.

I would speculate that the readiness of many climate skeptics to assume that climate scientists are motivated by greed reflects, at least in part, a failure of the moral imagination. Climate skeptics are predominantly conservative and libertarian: champions of the private sector, markets, the individual's unfettered pursuit of happiness, and the profit motive. To paint with a broad brush, the typical skeptic who extrapolates from his own personal and cultural experience may find the lure of material rewards a probable default explanation for others' behavior.

If I am right about this, science communication that paints an accurate portrait of individual climate scientists as persons might help the public form a more accurate estimation of the scientists' character and motivations, and might help climate skeptics appreciate and understand a personality type that is significantly different than their own.

By the same token, environmental activists—the sort of persons who are inclined to act in solidarity to limit their own standard of living for the sake of future generations—might make a default assumption that climate scientists conform to a similarly self-abnegating, heroically public-spirited stereotype. Some may, but others certainly do not. This too could be clarified with greater knowledge of actual scientists' personalities.

Journalist Fred Pearce (2010) rightly describes the circumstances that led up to the Climategate affair (the hacking of computers and release of thousands of Climatic Research Unit documents) as a "tragedy of misunderstood motives" (p. 13). The climate scientists assumed that skeptics were simply out to harass and disrupt legitimate science, not to acquire knowledge. Skeptics assumed that the scientists' refusal to share data meant the scientists had something to hide. In each case, a more accurate estimation of the others as persons—their knowledge, experience, habits of thought—might, arguably, have defused the conflict.

Ethical Guidelines for Communicating about Persons

In this section of the chapter, I introduce and discuss two broad and multifaceted ethical imperatives for communication about persons (in general and in the context of science). One is addressed to the rights of the reader, the other to the rights of the subject.

Many of the ethical insights discussed in this section come from the study of biography. I will not argue in detail in the case of each insight, but I assert here—and invite the reader to consider for him- or herself whether it is true—that what is true for biography also applies, mutatis mutandis, to other genres under consideration in this essay (journalism, social science, and the more implicit forms of personal characterization and self-presentation).

FIRST IMPERATIVE (VIS-à-VIS THE READER): BE TRUTHFUL

What does the writer owe the reader? The first, most important, and most universally acknowledged ethical rule for communicating about persons is the imperative to be truthful (e.g., Eakin, 2001, pp. 113–114; Eakin, 2004, p. 1ff; Hankins, 1979, pp. 1–2; Lockridge, 1999, p. 133; Mills, 2004, p. 104; Shortland & Yeo, 1996, p. 35).

This simple imperative is anything but simple to put into practice. As historian of science Thomas Hankins (1979) observes, it "places a moral obligation upon the biographer, but does not give him much direction" (p. 1). What does it mean to be truthful when talking or writing about a person's character, and can truthfulness be achieved in absolute terms?

Fact versus essence. The imperative to be truthful can be understood as an imperative to be factual. This means avoiding outright falsehoods and also eschewing narratively convenient fictions like invented dialogue (Lockridge, 1999, p. 133).

But even strictly hewing to facts is not enough. Biography should be, in Hankins's (1979) words (epitomizing the view of eminent biographers such as Nicolson and Condorcet), "both a true *and essential* account—one that not only relates facts, but also uses those facts to recreate accurately the subject's character" (pp. 1–2, emphasis added).

As it is possible to lie with statistics, so it is with facts. In a qualitative analysis of any kind presented in prose, a general point is usually illustrated with no more than a handful of examples. In the life-writing context, it can hardly be doubted that a handful of (true) examples of behavior could be cherry-picked to support nearly any conceivable generalization about an individual's character. It is wrong to cherry-pick examples of behavior to fit a preconceived notion of character, and it is also wrong to generalize (without appropriate caveats) about character from a limited sample of observations. Reliable assessments of character will be based on extensive and intimate knowledge. Life writing is thus "an inductive process, the biographer building his character as much as possible from the evidence, and not from any preconceived ideas or anachronistic interpretations" (Hankins, 1979, p. 2), and in this sense follows methods and norms not unlike those commonly prescribed for science.

Imposition of the author's perspective. Any writing necessarily imposes some sort of order on a complex and subtle reality. Lockridge (1999) observes that "biographers are always confronting narrative choices based on their sense of probability" (p. 132). In other words, the form of the biography is under-determined by the facts of the life, and the biographer must make choices and judgments, choices that cannot fail to reflect the biographer's own worldview, maturity, and character.

This is an instance where biography is also autobiography. Science historian Mary Jo Nye (2006) points to the example of Richard S. Westfall, who "arrived at the insight that the Puritan ethic that informed his own life furnished the set of categories that he used to construct his picture of Newton" (p. 327). Westfall frankly acknowledged that "in writing [Newton's] biography I have nevertheless composed my own autobiography . . . a portrait of my ideal self, of the self I would like to be" (as cited in Nye, 2006, p. 327).

Westfall's predicament may be inescapable in principle, but a life writer should be on guard against it and resist it. As we observed earlier, another's character and worldview may be radically different than our own. Lockridge (1999) urges us to respect the "alterity" (a Coleridgean term) of the other (p. 129). If we contentedly interpret the life and character of another as merely a distorted mirror of our own, we will fail to do justice to the subject and fall short of our responsibility to the reader.

In an essay about science biography, Söderqvist (1996) suggests how the problem of alterity may be tackled. The writer must "adopt an empathetic stance which does not falsify the scientist's position by imposing an alien vocabulary." He or she must "be sensitive to the vocabulary the person uses about himself, his work and the world around him"; in other words, the writer must "go native" (p. 63). To be sure, we can never step outside our own personal perspective and mental categories. But in studying another human being with empathy we can strive to expand our perspective and augment our set of mental categories, what Söderqvist calls our vocabulary.

Emotional involvement. As Medawar observed, every scientist has some emotional involvement with his or her chosen topic, be it ants, or plants, or planets. This involvement makes absolute objectivity unattainable. The biographer, too, necessarily has emotional involvement with his or her subject. The fact that the subject is a human being—and sometimes a living human being—makes the problem of emotional involvement more acute and complex.

As Söderqvist (2003, p. xxi) and Smocovitis (2007, pp. 216–217, see also literature cited on p. 209) testify, those who write about scientists are susceptible to attachment, identification, and emotional involvement. The emotional involvement need not be sympathetic. Biographers may come to loathe their subjects. Nye (2006, p. 328) lists Thomas Hager (biographer of Linus Pauling), Söderqvist (biographer of Jerne), and possibly Martin Sherwin (co-biographer of Robert Oppenheimer) as examples of disaffected scientific biographers. Söderqvist (2003) records that at one point he needed to unburden himself to a therapist who specializes in treating biographers (!), and it was only by giving himself permission "not to like Jerne" that he was able to complete the biographical project (p. xxi).

Personal feelings add an irreducible bias to communication about persons. They make perfect objectivity impossible. But they do not make the exercise entirely futile. It is possible to strive for objectivity and detachment and to achieve it to a substantial degree. As Lockridge (1999) puts it, "all biographers have an interest of one sort or another. . . . But the standard of disinterestedness remains to help guard against dual errors—the negative idealization of the hatchet job, and the positive of hagiography" (p. 133).

Multiple biographers report that in spite of strong emotional involvement with the aging rock-star scientists they interviewed and shadowed, they were able to achieve substantial detachment after the death of the subject (Smocovitis, 2007, p. 217; Söderqvist, 2003, p. xxi). Death, of course, is not required: the key ingredient is personal distance and the passage of time, which puts all sorts of relationships, not just those between biographer and subject, in proper perspective. And the process of writing itself may help to achieve the necessary emotional distance (Söderqvist, 2003, p. xxi–xxii). "The final result should emerge as a happy divorce," testifies Söderqvist (2003), "a certification that the writer has freed himself from the central figure" (p. xxii).

Emotional involvement poses a challenge to the biographer, but adds to the depth of insight the biographer is able to achieve. Smocovitis (2007), biographer of evolutionary biologist George Ledyard Stebbins, reflects that "conflicting feelings . . . if explored and integrated into the writing of the biography, may even enhance the quality of the biographical product" (p. 216). They make possible a portrait that is more vivid, more "true to life" (Smocovitis, 2007, p. 217).

Striving for objectivity—a reflection on the writer's own character. Ultimately, what truthfulness requires is a *commitment to accuracy and objectivity* on the part of the writer in spite of epistemic challenges and ambiguities and personal feelings. This speaks to the character of the writer, whether a biographer, journalist (cf. Nelkin, 1987, p. 91), or social scientist. As Lockridge

(1999) says, "Even if truth-telling . . . cannot be absolutely commanded in either biography or autobiography," it is "*meritorious* wherever and to the degree this happens" (p. 139, emphasis added).

Truthfulness in first-person writing. Both the biographer and the autobiographer must meet minimum standards for factual accuracy. But beyond that, there is a certain asymmetry that leads us to allow a first-person writer greater leeway (see Lockridge, 1999, p. 136ff).

In the first place, one simply can't be as objective about oneself as others can. We have unmatched access to the data of our own lives, but we simply do not have the opportunity to cultivate the detachment necessary for a balanced account.

That much is methodological. The second reason is more essential. Writing in the first person is not simply describing or narrating, it is active self-fashioning: "The autobiographer's act of composition may be regarded as itself part of a life still in the making. Writing autobiography is a continuation by other means of the life being narrated, on the edge of the compositional present" (Lockridge, 1999, p. 138). Personal autonomy requires that we respect an individual's right to explore self-definitions and find meaning on their own terms. That does not mean an outside observer can't pass judgment on the success of the autobiographical effort, and it doesn't excuse the autobiographer from striving for truth, or give the autobiographer the latitude to harm others in his or her writing. But it suggests a wider range of tolerance than we would give to a third party. In the words of Lockridge (1999), "if the autobiographer is on oath, it is to construct an identity plausible and public enough for the reader to recognize it, to entertain it. . . . So considered, a 'self-mythology' may be the autobiographer's truth and a usable public truth as well" (p. 139). A sincere commitment to the goals and standards of a scientific profession as a life project may be considered such a "self-mythology."

SECOND IMPERATIVE (VIS-à-VIS THE SUBJECT): BE RESPECTFUL

Both the scientist (student of things) and the biographer (student of persons) have an obligation to be truthful. But the biographer, whose subject is a person's life, livelihood, and character, has an additional obligation: to be respectful. Being respectful doesn't mean refraining from criticism, even strong criticism, but it sets some bounds to life writing.

Judicious exercise of biographer's "predatory" power. The civilized urban carnivore may seek, by taking its carcasses in the form of breaded nuggets and hamburger patties, to escape from the guilty knowledge that he or she is responsible for taking lives. The indigenous hunter, on the other hand, recognizes and accepts his or her culpability and seeks to expiate or redeem it.

The biographer too is a carnivore of sorts. Legal scholar Jeffrey Rosen (as cited in Eakin, 2004, p. 8) observes that "there are few acts more aggressive than describing someone else." Journalist Janet Malcolm (as cited in Eakin, 2004, p. 9) characterizes the biographer as a "professional burglar, breaking into a house, rifling through certain drawers that he has good reason to think contain the jewelry and money, and triumphantly bearing his loot away." The burglary metaphor is only barely an exaggeration. Quite actual "rummaging through closets and reading of other people's mail" is all in a good day's work for the biographer (Lockridge, 1999, p. 132).

Lockridge (1999) states frankly that "the biographer incurs the moral jeopardy of a predator":

> In a sense, the biographer takes possession of another, in the current metaphors colonizing or cannibalizing the life in *finding it out*, and using the subject as a means, especially the personal career advantage of getting a book published. This violates Kant's imperative to treat other human beings not as means but as ends in themselves. (p. 132)

This violation can be keenly felt. One of my early scholarly articles, which dissected the rhetoric of the public climate debates, was in turn reviewed and dissected on several prominent climate blogs. As a fairly private person, I found this attention partly flattering, but also discomfiting. My article argued, in part, that when we pass judgment on the credibility of others' arguments and conclusions, frequently we also pass judgment on their characters as well. In the ensuing blog chatter, I unsurprisingly found myself on the dissecting table alongside my own arguments and conclusions. I was called "naive." It was assumed that I was the sort of person who "can't do science" and therefore turns to social science "to create an impression of relevance." Though I was "brainy" and "an apparently reasonable, even-handed, intelligent figure," the fact that I had not grasped "the blindingly simple truth" about X and Y and Z made it clear that I was "obviously disingenuous." One commenter objected to my "prissy writing style" and "the apparently self-conscious attempt to appear above the fray." I was accused of wasting time and probably taxpayer money. I was given props for allegedly "admitting" that my own article was a "weak effort."

Lockridge (1999) speaks the discomfort we feel "when overhearing others speak of us—our indignation when they have it wrong, our despair when they have it right" (p. 133). That was my experience.

But if I was a victim, I was also as guilty as any predator. The bloggers who judged me did so on the basis of words that I had carefully polished and submitted to a journal for publication, in full knowledge (or hopeful expectation) that they would be read by strangers. I, on the other hand, based my analysis in part on the stolen private emails of climate scientists.

In other words, I myself had trespassed on other's lives (in fact, both the beleaguered scientists' and their opponents') by describing them in my words, and by treating their own public and private communications as artifacts or, in Lockridge's (1999) terms, by examining "the debris of somebody else's life" (p. 132). Where is justice in this scenario? Can such an offense be justified or redeemed?

To state the problem more generally: as the predator has no other way to subsist but to eat flesh, the scholar or journalist who wants to communicate something of value about science and scientists has no other choice but to trespass against scientists as persons. As Lockridge (1999) says of the conflict between the interest of the biographer and the rights of the subject, to categorically decide the "irresolvable moral antinomy . . . in favor of the subject would be to pre-empt the entire biographical project" (p. 132). Ethical philosopher Claudia Mills agrees (as paraphrased and quoted by Eakin, 2004, p. 10) that "the cost of telling someone else's story is inescapable: 'The sharing of stories does require that stories be shared.'"

Given that the writer who writes about people will trespass morally on his or her subjects, what is to be done? The trespass must be expiated with, as Mills (2004) writes, "appropriate care and respect for the stories told . . . with sensitivity and concern both for the stories themselves and even more for the persons, for the human beings, whose stories these are" (p. 114). What, then, are the relevant criteria? What exactly does "appropriate care and respect" entail?

Telling stories with redeeming value. First, one must ask: Is this a story that deserves to be told? Does it have some redeeming value? Mills (as paraphrased by Eakin, 2004, p. 10) argues that "to say that one is simply telling the truth, the usual all-purpose defense embraced by life writers whose motives have been impugned, is not enough to let one off the hook." Literary scholar Nancy K. Miller (as paraphrased by Eakin, 2004, p. 11) also finds that "telling the truth is not . . . in and of itself enough to justify disclosure."

Justifications for putting another person under the microscope range widely. Most scholars would agree that making a "contribution to knowledge" is normally sufficient. Journalists may appeal to the public interest. Many would agree with Mills (paraphrased and quoted in Eakin, 2004, p. 10) that voyeurism, "the debasement of 'talk show broadcasting,'" is not sufficient.

The difference between a story with redeeming value and one without is not only found in the subject matter: The manner of telling the story matters just as much, or more so. So criticism or a harsh assessment of another person, for example, can be character assassination or can be something more edifying. Mills (2004) quotes Ueland's praise of the "honesty, earnestness, and extraordinarily clear vision" of the great Russian writers in this regard:

> When they write about repulsive people, whom no doubt they knew well, there is nothing caddish or reprehensible about it. . . . Why is that? Is it because Tolstoi and Chekhov and Dostoyevsky and Gorky were so serious, so impassioned, so truthful about everything and would never let themselves show off or jeer or exaggerate? If you are serious in describing bad people and not mean or derisive or superior . . . even the bad people will be grateful. I would never resent being described by Chekhov, no matter how repellent the picture. I would try to be better. If Sinclair Lewis did it, or D. H. Lawrence or H. L. Mencken I would sue for libel. (p. 114)

This is another instance where the (perceived) character of the writer matters. Earnest or snide can make the difference between a worthy and unworthy story, and one that will fall on receptive or deaf ears.

Extra vigilance in truthfulness, out of respect for the subject. Second, respect for the subject means a redoubled commitment to truthfulness. This is because errors and failures of judgment not only reflect poorly on the character of the writer (if they are found out), but also may easily have repercussions for the subject or his or her memory. As Lockridge (1999) writes: "Publishing a biography inevitably poses the question of the subject's moral worth within the virtual court of a readership, whether or not the biographer encourages such judgment" (p. 133). Everyone has a right to a fair trial. When a writer drags another person into the court of public opinion, the writer has an obligation to be scrupulous about the facts and what can or cannot be inferred from them.

Respecting boundaries and rights of the subject. Respect also means setting and observing reasonable boundaries, so as to respect privacy and not to cause undue harm. Söderqvist (2003), for example, reports that in his biography of Jerne he refrained from divulging any more details of the intimate personal life of the subject than necessary to make points essential to the narrative, and wrote as little as possible about living family members (p. xxi). Smocovitis (2007) similarly felt obligated, as a guest invited into the homes of her subject and his wife, to take seriously the "trust, confidence and the dignity that both had a right to" (p. 214).

What constitute "reasonable" boundaries can vary widely from genre to genre, and reasonable people might disagree. Greater latitude is forgivable with dead subjects than with living ones.

Striving for appropriate care and respect—a reflection on the writer's own character. Each of these criteria for respectfulness—telling a worthy story, striving for truthfulness, and seeking to minimize harm—can be seen as a reflection on the character of the author. As Lockridge (1999) says, capturing at least two of the three criteria, "by way of compensation for violation of the subject, the biographer may aspire to certain virtues—of disinterestedness, honesty, accuracy and fairness in a just telling of the life, a life that for one reason or another *ought* to be told, after all" (p. 133).

Conclusion

In closing, it is worth remarking again on the parallels between common ideas about the ethic of science and the ethics of writing about science. Another way of saying this, as I have remarked in other contexts (Ranalli, 2012a, p. 200), is that scientific norms shade neatly into general scholarly norms.

Both the scientist and the science writer are expected to demonstrate a commitment to objectivity and truthfulness—a commitment that is no less crucial for the fact that its object can never be achieved in absolute terms. Both are expected to strive for emotional detachment, despite their very human interest in and involvement with their subject.

Less obviously and less transparently, both the scientist and the science writer are engaged in self-fashioning at every turn. They are aiming to demonstrate competence and credibility, among other virtues. And in striving to excel at their craft, they are answering the moral question, what is it good to be?

The imperative to be respectful is one area where the parallel between the scientist and the science writer (or other humanist) appears to break down. In the physical sciences at least, we do not generally expect scientists to recognize ethical obligations toward their objects of study. Many of us would find it difficult even to conceive how one might incur ethical obligations toward quarks and quasars. But the prospect is worth considering. The modern, secular view of the world as composed of inert matter is, after all, historically aberrant (Liess, 1972; Merchant, 1980, 1989). Before the emergence of the modern scientific worldview in Europe in the 1600s, most cultures, even monotheistic cultures, viewed the world as (to a large extent) "animated," populated with ensouled objects. It would be an error to romanticize indigenous and pre-modern animistic worldviews, but on the other hand historian of science and cultural critic Morris Berman (1981) makes an intriguing case that the modern "disenchanted" way of viewing and treating the world as inanimate is so extreme as to be pathological. Berman suggests that a more balanced, humane synthesis would involve treating the world and its many components as to a certain degree animate—imbued with a certain degree of personhood that commands respect.[5]

Henry David Thoreau is sometimes held up as an exemplar of a more humanized scientific practice, one that engages the scientist as a whole person and seeks to treat the scientific object as much as possible as a subject (Walls, 1995). His famous poetic sensibility gave him empathy with both living and nonliving objects of study. (The story is told that when a neighbor asked Thoreau if it was true he didn't shoot birds in order to study them—as did contemporaries like James Audubon and Louis Agassiz—he retorted, "Do you think that I should shoot you if I wanted to study you?"; Harding, 1982, p. 356.)

Thoreau's poetic sensibility did not, as one might be tempted to fear, preclude the emotional distance required for objectivity: Thoreau equally famously possessed a ruthless critical sensibility (honed, arguably, by his practice of journaling) that he applied generously to himself and to the habits and opinions of others. The poetic sensibility (anthropomorphizing the owls, loons, and whip-poor-wills around Walden Pond or seeing armies of Troy and Greece in ant colonies) did not impede in the least his ability to produce valid findings in phenology, ethology, ecology, etc. To speak more generally: a capacity for empathy and a capacity for skeptical inquiry might seem to be opposed, but they are actually two separate virtues and it is possible to cultivate both. As Smocovitis (2007) pointed out in the context of scientific biography, to write well about a subject requires both that we achieve sufficient closeness to generate insights, and that we subsequently achieve sufficient distance to process and frame those insights objectively.

Scientists today follow basic ethical guidelines for the treatment of human and animal subjects. Might they profitably follow Thoreau's example even farther in this direction? The exercise we have undertaken in this essay suggests that, should scientific practitioners aim to adopt or adapt it, a generalized ethic of respectfulness toward objects of study is available from humanistic scholarship.

---

5. On personhood as a complex concept that admits of degrees, see Bird-David (1999), Harvey (2005), Ranalli (2012b).

References

Berman, M. (1981). *The reenchantment of the world*. Ithaca, NY: Cornell University Press.

Bird-David, N. 1999. "Animism" revisited: Personhood, environment, and relational epistemology. *Current Anthropology, 40*, S67–S91.

Eakin, P. J. (2001). Breaking rules: The consequences of self-narration. *Biography, 24*(1), 113–127.

Eakin, P. J. (2004). *The ethics of life writing*. Ithaca, NY: Cornell University Press.

Goffman, E. (1959). *The presentation of self in everyday life*. New York, NY: Doubleday Anchor.

Gross, A. G., Harmon, J. E., & Reidy, M. S. (2002). *Communicating science: The scientific article from the 17th century to the present*. Oxford, UK: Oxford University Press.

Haack, S. (1993). *Evidence and inquiry*. Oxford, UK: Blackwell.

Hankins, T. L. (1979). In defence of biography: The use of biography in the history of science. *History of Science, 17*(1), 1–6.

Harding, W. (1982). *The days of Henry Thoreau: A biography*. Princeton, NJ: Princeton University Press.

Harvey, G. (2005). *Animism*. London: Hurst.

Kahan, D. M., and Braman, D. (2006). Cultural cognition and public policy, *Yale Law & Policy Review, 24*(1).

Liess, W. (1972). *The domination of nature*. New York, NY: George Braziller.

Lockridge, L. (1999). The ethics of biography and autobiography. In D. Rainsford & T. Woods (Eds.), *Critical ethics: Text, theory and responsibility* (pp. 125–140). New York, NY: St. Martin's Press.

MacIntyre, A. C. (1978). Objectivity in morality and objectivity in science. In H. T. Engelhardt Jr. & D. Callahan (Eds.), *Morals, science, and sociality* (pp. 21–39). New York, NY: Institute of Society, Ethics, and the Life Sciences.

MacIntyre, A. C. (1984). *After virtue: A study in moral theory*. Notre Dame, IN: University of Notre Dame Press.

Medawar, P. B. (1963). Is the scientific paper a fraud? *Listener, 70*(12), 377–378.

Merchant, C. (1980). *The death of nature: Women, ecology, and the scientific revolution*. San Francisco, CA: Harper & Row.

Merchant, C. (1989). *Ecological revolutions: Nature, gender, and science in New England*. Chapel Hill: University of North Carolina Press.

Merton, R. K., & Storer, N. W. (1973). *The sociology of science: Theoretical and empirical investigations*. Chicago, IL: University of Chicago Press.

Mills, C. (2004). Friendship, fiction, and memoir: Trust and betrayal in writing from one's own life. In P. J. Eakin (Ed.), *The ethics of life writing* (pp. 101–120). Ithaca, NY: Cornell University Press.

Nelkin, D. (1987). *Selling science: How the press covers science and technology*. New York, NY: W. H. Freeman.

Nye, M. J. (2006). Scientific biography: History of science by another means? *Isis, 97*(2), 322–329.

Parker, D. (2004). Life writing as narrative of the good: Father and Son and the ethics of authenticity. In P. J. Eakin (Ed.), *The ethics of life writing* (pp. 53–72). Ithaca, NY: Cornell University Press.

Pearce, F. (2010). *The climate files: The battle for the truth about global warming*. London, UK: Guardian Books.

Ranalli, B. (2012a). Climate science, character, and the "hard-won" consensus. *Kennedy Institute of Ethics Journal, 22*(2), 183–210.

Ranalli, B. (2012b). Three grand narratives: Historical links between forms of economic life and religion. *Études maritainiennes—Maritain Studies, 28,* 33–46.

Shapin, S. (2010). *Never pure: Historical studies of science as if it was produced by people with bodies, situated in time, space, culture, and society, and struggling for credibility and authority.* Baltimore, MD: Johns Hopkins University Press.

Shortland, M., and Yeo, R. (1996). *Telling lives in science: Essays on scientific biography.* Cambridge, UK: Cambridge University Press.

Smocovitis, V. B. (2007). Pas de deux: The biographer and the living biographical subject. In T. Söderqvist (Ed.), *The history and poetics of scientific biography* (pp. 207–220). Burlington, VT: Ashgate.

Söderqvist, T. (1996). Existential projects and existential choice in science: Science biography as an edifying genre. In M. Shortland and R. Yeo (Eds.). *Telling lives in science: Essays on scientific biography* (pp. 45–84). Cambridge, UK: Cambridge University Press.

Söderqvist, T. (1997). Virtue ethics and the historiography of science. *Danish Yearbook of Philosophy, 32,* 45–64.

Söderqvist, T. (2003). *Science as autobiography: The troubled life of Niels Jerne.* D. M. Paul (Trans.). New Haven, CT: Yale University Press.

Söderqvist, T. (2007). *The history and poetics of scientific biography.* Burlington, VT: Ashgate.

Thoreau, H. D. (1971). *Walden.* J. L. Shanley (Ed.). Princeton, NJ: Princeton University Press. (Original work published 1895)

Walls, L. D. (1995). *Seeing new worlds: Henry David Thoreau and nineteenth-century natural science.* Madison: University of Wisconsin Press.

Watts, A. (2011, March 6). Briggs on Berkeley's forthcoming BEST surface temperature record, plus my thoughts from my visit there [Weblog post]. *Watts Up With That?* Retrieved from http://wattsupwiththat.com/2011/03/06/briggs-on-berkeleys-best-plus-my-thoughts-from-my-visit-there/

Ziman, J. (2002). *Real science: What it is, and what it means.* Cambridge, UK: Cambridge University Press.

7

Journalists, Expert Sources, and Ethical Issues in Science Communication

MARJORIE KRUVAND

Americans who follow science or medical news have likely seen, heard, or read about Arthur L. Caplan, professor and founding head of the Division of Medical Ethics at New York University School of Medicine. For more than two decades, he has been a ubiquitous expert source on a broad range of topics: weighing in on the ethics of assisted suicide, opining on test kits that enable pregnant couples to learn their baby's sex at home, arguing for greater protections for human research subjects, predicting whether consumers will eat meat from cloned animals, contending that a 17-year-old girl can't refuse chemotherapy she says she doesn't want, and criticizing parents who refuse to vaccinate their healthy children. Caplan is interviewed frequently on television and quoted in scores of newspapers, magazines, and online news sites each year. He "basks in attention from reporters who, on a day when a big medical story breaks, instinctively telephone him, knowing that he will quickly return that call with a useful quote and appealing observation" (Meier & Thomas, 2015).

Caplan is an example of the relatively small number of expert sources who have long dominated US media coverage of science and medicine. More than 40 years ago, Rae Goodell coined the term "visible scientists" to describe B. F. Skinner, Isaac Asimov, Margaret Mead, Paul Ehrlich, Linus Pauling, and James Watson, who shared a rare ability in the scientific community: Communicating effectively with reporters and the public and garnering headlines in the process (1977). More recently, visible scientists have included planetary scientist Carl Sagan, whose frequent media appearances helped increase public understanding of, and support for, science; biologist Craig Venter, who

This chapter has been updated and modified from an article previously published in *Science Communication: Linking Theory and Practice*, 34(5): 566–591, October 2012.

has become a go-to resource to describe the complex technology of synthetic biology; and astrophysicist Neil deGrasse Tyson, whose clear and engaging scientific explanations, coupled with his enthusiasm, have made him a favorite of journalists and the public alike.

But repeated reliance on a limited number of experts may pose ethical issues for both reporters and their sources. That's because the experts journalists select as sources influence how news of science and medicine is presented to the public. Expert sources can play inconspicuous but powerful roles as news shapers and opinion influencers (Conrad, 1999; Soley, 1994). This practice has greater significance today because hundreds of experienced science and medical reporters have been among the tens of thousands of US journalists who have been laid off or have taken buyouts in the past decade (Petit, 2007; Russell, 2006; Tenore, 2009). If they are replaced at all, it is typically with less-experienced reporters assigned to cover multiple topics rather than specialize in science and medical reporting. In addition, journalists today are often required to produce several versions of the same story for different platforms, such as a newspaper and a website or a television newscast and Facebook, while meeting 24/7 news deadlines:

> Replacing the "old timers" for a digitally-oriented, younger and cheaper staff is an easy tactic to employ, but not necessarily a smart one. It deprives audiences of veteran journalists' experience and reinforces the idea that quantification (more clicks, always) reigns over the quality of content (Pinheiro, 2015).

These changes in US newsrooms limit the amount of research journalists can do and the number of sources they can contact. So journalists are likely to rely more than ever on a small cadre of trusted sources to help them be efficient and productive. As a result, the dominance of a relatively few expert sources in science and medical stories is likely to continue.

While this chapter explores this topic through the lens of a single expert source, Caplan, it also has implications for other experts on whom journalists depend. The news media is a critical vehicle, and one of the most accessible, for communicating scientific and medical information (Conrad, 1999), as well as the primary arena in which scientific issues come to the attention of decision makers, interest groups, and the public (Nisbet, Brossard, & Kroepsch, 2003). Members of the news media serve as brokers between science and the public,

> framing social relationships for their readers and shaping the public consciousness about science-related events. . . . Through their selection of news, journalists help set the agenda for public policy . . . and through the presentation of science news, the media influence public attitudes towards science. (Nelkin, 2001, p. 205)

How Journalists Work

Reporters follow organizational routines that have a significant influence not only on the ways in which they work, but also on the shaping of news content (Shoemaker & Reese, 1996; Tuchman, 1978). For example, journalists look for convenient, timely stories that align with their deadlines (Shoemaker & Reese, 1996), strive for a balance among competing views and objectivity in their reporting (Kovach & Rosenstiel, 2001), and rely heavily on sources. These news routines are useful in examining how reporters develop stories as well as understanding the influences on those processes and on news media content.

Journalists depend on others for much of the content in their stories (Conrad, 1999). A reporter, "especially when under deadline pressure and dealing with a complex story, is a captive of his or her sources" (Rubin & Hendy, 1977, p. 772). By providing information, context, and opinion that define and shape the story, a reporter's choice of sources "powerfully influences how that story is told" (Maier & Kasoma, 2005, p. 1). Sources may be proactive, initiating contact with reporters—either themselves or through public relations practitioners—or wait for a reporter to call (Gans, 1979; Reich, 2006).

Reporters work most efficiently when they know in advance what sources they plan to interview will say (Shoemaker & Reese, 1996). Journalists therefore develop a relatively small roster of trusted sources they know will provide certain information or opinion needed to flesh out a story (Conrad, 1999; Shoemaker & Reese, 1996; Reese & Danielian, 1994). Reporters "find it easier and more predictable to consult a narrow range of experts than to call on new ones each time" (Shoemaker & Reese, 1996, p. 131). As a result, "eager sources eventually become regular ones, appearing in the news over and over again" (Gans, 1979, p. 118).

Journalists and Expert Sources

Certain types of sources, including government officials, scientists, physicians, and industry officials, are much more prevalent in stories on science and medicine than other types of sources (Conrad, 1999; Dunwoody, 1987; Dunwoody & Ryan, 1987; Goodell, 1977; Nelkin, 1995; Nisbet & Lewenstein, 2002; Sumpter & Garner, 2007). Journalists rely on experts to help explain and interpret events in hopes of enhancing objectivity, credibility, and authority (Boyce, 2002; Steele, 1995); in turn, the experts become news shapers by providing comment and context for stories (Conrad, 1999; Soley, 1994).

Journalists need expert sources for information, explanation, context, implications, and opinion. But what journalists really *want* from experts is accessibility; exclusivity; rapid response; quotes that are pithy, colorful, and memorable; and comments that promote controversy and sensationalism (Albaek, Christiansen, & Togeby, 2003; Boyce, 2006; Kruvand, 2012). In other words, journalists would prefer that scientists and physicians communicate in ways antithetical to their professional training. That's because of the differences in the fields of science, medicine, and journalism and the missions of scientists, physicians, and journalists. While science and medicine are "slow, precise, careful, conservative and complicated," journalism is "hungry for headlines and drama, fast, short, dramatic, (and) very imprecise at times" (Sawyer, cited in Hartz & Chappell, 1997, p. 14). Scientists and physicians care greatly about context, precision, qualification, and nuance (Gregory & Miller, 1998). In contrast, journalists care about timeliness, accuracy, balance, and, above all, getting a story—the bigger and juicier the better (Dentzer, 2009).

Although the use of expertise in society is on the rise, public trust in experts has declined (Boyce, 2006; Limoges, 1993). As one communication scholar notes, "We believe less and less in experts . . . [but] we use them more and more" (Limoges, 1993, p. 424). Nearly twice as many experts were quoted in three US newspapers in 1990 as in 1978 (Soley, 1994). And a study of Danish newspapers found that the number of expert sources increased dramatically over the past 40 years (Albaek et al., 2003). Greater media competition and declining levels of public trust in journalism have prompted increased use of expert sources (Albaek et al., 2003). Including more experts may highlight conflict and tension in stories, making them more interesting and relevant. Another factor is the growing complexity of the news; 60 percent of local television health reporters surveyed said they must frequently find a health expert to explain complicated information because of the technical nature of medical news (Tanner, 2004).

There has been limited research on who expert sources are and how they are used (Conrad, 1999; Kruvand, 2009; Soley, 1994; Steele, 1995; Sumpter & Garner, 2007), as well as how reporters evaluate their expertise (Boyce, 2006; Kruvand, 2012; Martin, 1991; Soley, 1994; Steele, 1995; Stocking, 1985). But selecting expert sources is more challenging than choosing other types of news sources: How can a non-expert identify a genuine expert, especially in an obscure or highly specialized field? (Boyce, 2006; K. W. Goodman, 1999). As columnist Ellen Goodman notes, "Every reporter worth his or her Rolodex has a list of duly and not necessarily legitimately dubbed experts. One of the most ludicrous phrases in modern journalism has become, 'Experts say . . .'" (1997).

Criteria journalists use to select expert sources include credentials, qualifications, reputation, accessibility, efficiency, reliability, and prior media visibility (Boyce, 2006; Conrad, 1999; Friedman, Dunwoody & Rogers, 1986; Gans, 1979; Kruvand, 2012; Stocking, 1985; Van Dijk, 2004). Reporters covering science and medicine also value expert sources with highly visible names, titles, and affiliations, and even a touch of celebrity (Conrad, 1999; Goodell, 1977; Shepherd, 1981). But in science and medical reporting, expertise is not enough; sources must also be able to avoid technical jargon and explain information or provide opinion in plain English (Burkett, 1986).

Why Sources Matter

Experts who talk with reporters frequently are better at saying things effectively for stories (Conrad, 1999). Some experts are willing co-conspirators in the newsgathering process, learning precisely what journalists want and giving it to them by "formulat[ing] message strategies to accent drama and familiar story formats" (Nisbet et al., 2003, p. 43–44). Therefore, the simplest thing for journalists to do, especially on deadline, is to call the same expert again and again (K. W. Goodman, 1999). As a result, some expert sources gain significant power to define the news (Brown, Bybee, Wearden, & Straughan, 1987).

Indeed, reporters on specialized beats can become co-opted by their sources (Gans, 1979). Because these reporters popularize technical knowledge, they can become very important to sources who value publicity (Gans, 1979). The mutually beneficial relationship between journalists and experts may be especially strong, and even detrimental, in science journalism (Goodell, 1986). Through "a particular kind of chauvinism," reporters and their sources often assume that the sources have definitive views based on their expertise (Goodell, 1986, p. 177). This may often lead to an uncritical and unwarranted boosterism of science (Goodell, 1986), or "selling science" to the public (Nelkin, 1995).

Reporters are trained to follow the journalistic norms of objectivity and balance by using experts with opposing viewpoints (Boyce, 2006; Conrad, 1999; Steele, 1995). But using experts on opposing sides can frame the parameters of debate on an issue (Evans, 2002). And journalistic balance is sometimes misconstrued by reporters as balancing two sides equally, giving the impression that public opinion is evenly divided although this may not be so (Kovach & Rosenstiel, 2001). Therefore, asking a small number of experts to make predictions, analyze motives, and provide commentary and analysis may undercut "the very goal of objectivity that encourages journalists to seek out experts in the first place" (Steele, 1995, p. 809).

Bioethicists and Reporters

Journalists covering science and medicine in the 1960s and 1970s chiefly used binary expert sources, contrasting the views of scientists (as advocates of scientific progress) and clergy members (as appointed moral guardians of society) on stories with ethical issues (Nisbet et al., 2003). But reporters covering similar stories today often include bioethicists as well (Kruvand, 2009; Nisbet et al., 2003). As bioethical issues have proliferated in science and medical news, bioethicists have emerged as "a new class of public expert" who "passes judgment on right and wrong, often on matters of life and death" (K. W. Goodman, 1999, p. 189).

Journalists have increasingly turned to bioethicists to help them make sense of the neologisms that have made their way into media headlines and public discourse: Test-tube babies, artificial hearts, surrogate mothers, persistent vegetative state, xenografts, Frankenfoods, the morning-after pill, and physician-assisted suicide are but a few examples. By the time Dolly the sheep was cloned in 1997, bioethics was firmly entrenched in the news media (Rosenfeld, 1999). As reporters courted bioethicists to help demystify cloning and its human consequences, bioethicists, once "obscure figures of academia, suddenly became recognized public figures" (Guthmann, 2006, p. E1).

Bioethicists acknowledge that the news media has bestowed public legitimacy on their profession (Simonson, 2002). Yet some bioethicists are uncomfortable with or disappointed by media coverage, contending that reporters "seem to simplify the complex, reduce deliberation and nuanced argument to sound bite, and favor the sensational over the carefully argued" (p. 32). Science and medical reporters face the challenge of converting "complex and ambiguous scientific findings into nontechnical, compelling, and readable stories" (Conrad, 2001, p. 91). The nuanced viewpoints of bioethicists can be lost in the process (Levine, 2007). There is also a gulf between the deliberateness of philosophy and the speed of journalism:

> Given their structure, rapid production schedules, and need to attract the attention of an audience that has a million other things to do, the mass media are necessarily fragmented, hurried, entertaining. They do have a very important function, but it is not the comprehensive, educational one demanded of them by political philosophers and disgruntled intellectuals. Against that standard they will always fail. Instead, their function is to set agendas or to

bring issues to our attention. And in that arena they succeed spectacularly (Dunwoody, 1987, p. 48).

Journalistic brevity can result in short quotes that pass moral judgment without supporting argumentation, producing "a cartoon of an ethical issue, not an account" (K. W. Goodman, 1999, p. 192):

> Reporters aren't interested in detailed analysis or lengthy qualifications. A short, pithy quote is what's wanted. Nor are reporters eager to hear reassurances that alarming events aren't alarming. That doesn't make good copy. What makes good copy is that the events being reported are morally troubling or worse (Rachels, 1991, p. 67).

Bioethicists may exacerbate the problem by making "snap judgments" (Rachels, 1991, p. 67) or "trying too hard to be pithy when an issue demands reflection" (K. W. Goodman, 1999, p. 194).

Although bioethicists may be regarded by the public and the media as interchangeable experts with a unified perspective (Tuhus-Dubrow, 2006), their academic training may be in medicine, philosophy, science, theology, or law. And there are Catholic bioethicists, Protestant bioethicists, Jewish bioethicists, feminist bioethicists, liberal bioethicists, conservative bioethicists, libertarian bioethicists, and communitarian bioethicists, to name a few categories. Each has a distinct background and worldview. Yet reporting on bioethical issues often seems to involve "seeking out the opinion of one 'bioethics expert' and presenting it at least tacitly as representing the views of all who are in this line of work" (K. W. Goodman, 1999, p. 193). But neither bioethicists nor their opinions are homogenous: While any biologist could explain a stem cell to a reporter in much the same way as their peers would do, no two bioethicists may provide exactly the same perspective on the ethical implications of human embryonic stem cell research.

Expert sources are expected to have authoritative opinions to inspire the confidence of journalists and news consumers (Mepham, 2005). But experts rarely admit they do not know the answer to questions in their field because that would imply incompetence. Thus the views of experts, "sometimes tentative and sometimes uninformed, tend to get vested with an authority which may not be justified" (Mepham, 2005, p. 326). Moreover, bioethicists may perpetuate the impression that there is only one possible correct moral position—theirs—by failing to emphasize that their views are personal (Pence, 1999). Because an expert source may become irked if a reporter seeks to balance what the expert sees as a definitive statement or comment (Boyce, 2006), a bioethicist may be hesitant to suggest other bioethicists with contrasting views: "The great danger . . . is that only a very few media-savvy bioethicists define to the public what 'bioethics' says about an issue" (Pence, 1999, p. 48).

Caplan's Franchise

Caplan is a well-regarded academic scholar who has developed a franchise as a frequent expert source on bioethical issues. But since there are 2,000 to 6,000 bioethicists in the United States (Ipaktchian, 2011), why is Caplan used as an expert source so frequently? How and why can a single expert source become so widely used by journalists that he or she becomes the de facto stand-in for an entire profession and threaten to dominate media discourse on certain topics? And what does this indicate about the state of science and medical reporting and the ways in which news consumers are exposed to ethical issues in science and medicine?

Caplan has a PhD in the history and philosophy of science from Columbia University. He was associate director of the Hastings Center, a bioethics institute in New York, from 1984 to 1987. He then taught at the University of Minnesota, University of Pittsburgh, Columbia University, and the University of Pennsylvania, where he also directed its Center for Bioethics. Caplan has written 35 books and received many awards and honors, including being named one of the 50 most influential people in American health care by *Modern Health Care* magazine and one of the ten most influential people in America in biotechnology by the *National Journal.* He has been described as

> The most well-known bioethicist in America and, perhaps, the world . . . In large part, this is because he is able to talk to the media in a way that everybody can understand, because he has the time to do so, and because he really wants to do so (Pence, 1999, p. 48).

To illuminate the use of bioethicists as expert sources in the media, stories on bioethical issues in six newspapers over a 19-year period (1992–2010) were analyzed. One objective was to examine when and how journalists use bioethicists as expert sources and which bioethicists they choose. The other was to investigate the extent of Caplan's use as an expert source. Coverage was drawn from *The New York Times, San Francisco Chronicle, The Atlanta JournalConstitution-, Houston Chronicle, St. Louis Post-Dispatch,* and *The Boston Globe. The New York Times* is an elite national newspaper whose prominence and authoritativeness also extend to science and medical reporting. The five regional newspapers had roughly comparable circulations[1] during the study period (Audit Bureau of Circulations, 2010). All six papers had science and/ or medical reporters during the study period.

A total of 1,136 stories were found: 46.7 percent from *The New York Times*, 21.6 percent from *The Boston Globe*, 10.2 percent from the *San Francisco Chronicle*, 8.0 percent from *The Atlanta Journal-Constitution*, 7.9 percent from the *Houston Chronicle*, and 5.6 percent from the *St. Louis Post-Dispatch*. From these a random sample of 548 stories was selected (for more on the study method, see Kruvand, 2012).

A total of 210 unique bioethicists were directly quoted in the stories analyzed. Of these, 123 (58.6 percent) appeared in a single story. Of the 87 bioethicists in more than one story, 43 were used as expert sources in one newspaper and 44 were used in more than one. Among the six newspapers, *The New York Times* used the highest number of unique bioethicists (137). A single bioethicist was directly quoted in 77.0 percent of the 548 stories analyzed. Two bioethicists were used as expert sources in 16.3 percent, three bioethicists appeared in 6.0 percent, and four or more in 0.7 percent.

Caplan was the only bioethicist directly quoted in all six newspapers. He was also used as an expert source most often; he was directly quoted in 188 stories, or 34.3 percent of the stories analyzed. Caplan appeared proportionately most often in *The Atlanta Journal-Constitution*; he was quoted directly in nearly two-thirds of the newspaper's stories (31 of 49 stories). George Annas, a bioethicist at Boston University, was in the second-largest number of stories—52, or less than a third as many as Caplan. After Annas, the number of stories in which a specific bioethicist was quoted declined sharply. Leon Kass, a professor emeritus at the University of Chicago and former chairman of the President's Council on Bioethics, was in 19 stories. Ten of the 11 bioethicists quoted most frequently were male. (See table 7.1.)

After the stories were analyzed, an hour-long phone interview was conducted with Caplan. In-depth phone interviews were also conducted with a present or former science or medical reporter at each of the six newspapers. Each reporter had bylined stories among the stories analyzed.

---

1. The *Houston Chronicle* had the largest average weekday circulation (343,952) in 2010 of the five regional newspapers. Three others had nearly identical average weekday circulations: the *San Francisco Chronicle* (223,549), *The Boston Globe* (222,683), and *St. Louis Post-Dispatch* (221,629). *The Atlanta Journal-Constitution* had the smallest average weekday circulation (181,504). All five papers have had substantial circulation declines since the beginning of the study period.

TABLE 7.1. Bioethicists Used Most Frequently as Expert Sources in Six US Newspapers, 1992-2010

Rank	Name	Affiliation during Study Period	No. of Stories	No. of Papers
1	Arthur Caplan	University of Pennsylvania	188	6
2	George Annas	Boston University	52	4
3	Leon Kass	University of Chicago	19	3
4	Norman Fost	University of Wisconsin	16	3
5	R. Alta Charo	University of Wisconsin	14	3
6	William Winslade	University of Houston	12	1
7 (tie)	Michael Grodin	Boston University	11	2
	David Magnus	Stanford University	11	2
8 (tie)	Daniel Callahan	The Hastings Center	10	3
	Bernard Lo	University of California, San Francisco	10	2
	Thomas H. Murray	The Hastings Center	10	2

How Journalists Select Bioethicists as Expert Sources

The reporters said the most important criteria in choosing bioethicists as expert sources were accessibility, responsiveness, reliability, and having something worthwhile to say. Alice Dembner, a former medical reporter at *The Boston Globe*, noted:

> It's no good to have an expert in your Rolodex if you can't reach them or they don't return your calls until next week. When you find an academic expert who is willing to operate on news deadlines rather than on academic deadlines and who gets what journalism is all about, I latch onto that person.

Dembner added that neither the name recognition of the bioethicist nor the prestige of the institution with which they are affiliated was essential. "The most important criterion is whether they have something intelligent and thoughtful to say," she said.

Carl T. Hall, a former science reporter at the *San Francisco Chronicle*, said reporters' need for efficiency drove his choice of bioethicists to interview. "I look for the person who's available with the most cogent, informed point of view," he said. "It's knowing that if you call them in a hurry, you'll probably get a usable quote. The ability to provide a good quote on deadline is a skill in itself."

Caplan said that when journalists call bioethicists, "they typically call the reporter back four days later, send them two papers to read first, and only talk in technical jargon." In contrast, Caplan clearly understands news routines, including the importance of translating technical issues into plain English.

He said that being available, getting back to journalists quickly, being concise, and having a sense of humor were attributes that helped him develop what he called a "big footprint" in the news media. "I know science and I think I can translate well," he said. "That's a skill also much appreciated by journalists."

The reporters interviewed said they value those characteristics in Caplan. "We called him 'Dr. Soundbite,'" said Maryn McKenna, a former medical reporter at *The Atlanta Journal-Constitution*. "He was incredibly media savvy and worked very hard at being accessible." Todd Ackerman, medical reporter at the *Houston Chronicle*, said he was impressed that Caplan agreed to be interviewed by cell phone while the bioethicist was in Norway. Dembner noted that Caplan "understands the time pressures reporters are working under." And Nicolas Wade, then a science reporter at *The New York Times*, said Caplan was very bright and very accessible, and "works incredibly hard."

The fact that Caplan does not talk like an academic or ethicist makes him even more popular with reporters. Wade called Caplan "an absolute master" of sharp, pithy, and colorful quotes. This is the case whether the topic is abortion: "What's depressing is to watch is the geography of politics cover the moral plate tectonics of abortion" (Toner, 2001, p. D1), cryonics: "[It's] goofy beyond amusement. It's a movement that combines . . . screwy science and a secular lust for reincarnation with large-scale refrigeration technology" (Foreman, 1993, p. 1A), or efforts to ban human cloning: "I think . . . these people have become susceptible to bogeyman nightmares about cuckoo scientists run amok" (Stolberg, 2002, p. D16). Some bioethics scholars have expressed concern about bioethicists "trying too hard to be pithy when an issue demands reflection" (K. W. Goodman, 1999, p. 194). But Caplan said "being pithy and concise is good in terms of what reporters need," adding, "joking and quipping are just part of my style."

Expert Overuse?

Over the past several decades, bioethicists have appeared frequently as expert sources in science and medical news. But the 210 unique bioethicists directly quoted in the 548 stories analyzed represent a small minority of the 2,000 to 6,000 bioethicists in the United States. A single bioethicist, Caplan, was used as an expert source in 188 stories. And the two most-quoted bioethicists, Caplan and George Annas of Boston University, were quoted in 240 stories, or 43.8 percent of the total. These findings support the view that "eager sources eventually become regular ones, appearing in the news over and over again" (Gans, 1979, p. 118). As a result, some expert sources accrue significant power to define the news (Brown et al., 1987). Caplan, who has been a reliable "go-to" source for journalists over the 19-year study period, clearly fits this description.

Caplan said he made a deliberate choice to work with reporters because of a longstanding interest in "pushing bioethics into the public arena." At first, Caplan acknowledged, it was "certainly a pleasure to get quoted or see your name out there." But by now, he added, it has "long ceased to impress friends or family members." Caplan said it has been a "tricky problem" in academic culture to be both a respected scholar and an expert source. He said he was wary of the "Carl Sagan syndrome," referring to the late astronomer and cosmologist who helped popularize science through the media but whose status as a serious scholar was questioned by other scientists. Nonetheless, Caplan said it was important to engage the public.

But is there a potential danger from reporters relying so heavily on so few bioethicists as expert sources? Some bioethicists criticize the close relationships between reporters and certain bioethicists (K. W. Goodman, 1999), worrying that "only a very few media-savvy bioethicists define to the public what 'bioethics' says about an issue" (Pence, 1999, p. 48). Caplan acknowledged that "sometimes my colleagues will get mad at me and say, 'we'd like to be heard some.'" But while Caplan believes that more bioethicists can and should talk with journalists, he added, "It's not for everyone. Not everyone is good at it."

McKenna said she and her former colleagues at *The Atlanta Journal-Constitution* eventually realized their overreliance on Caplan and informally agreed to limit his use. "Journalists have a responsibility not to overuse sources," she said. Wade said he tries to avoid using Caplan as an expert source because his colleagues at *The New York Times* use him frequently. Reporters should "try to cast the net wider and use other bioethicists, and that's what I do," he said. Dembner, formerly of *The Boston Globe*, noted: "Using the same bioethicist over and over leads to overload, yes. But whether I will find another bioethicist instead will depend on how much time I have."

None of the reporters said their editors ever raised concerns about repeated use of certain bioethicists as expert sources. Hall said quoting certain bioethicists habitually was no different than reporters using the same scientists, political officials, economists, retired military officials, or sports figures as expert sources again and again. Yet repeatedly asking the same small group of expert sources to provide analysis and commentary can undermine "the very goal of objectivity that encourages journalists to seek out experts in the first place" (Steele, 1995, p. 809).

Caplan was directly quoted on 104 bioethical issues in the 188 stories in which he was an expert source. The topics ranged from fetal testing to government oversight of clinical trials to plastic surgery for men with enlarged breasts. He also commented frequently on controversial local medical cases, such as when scarce or expensive resources or untested treatments were provided to, or withheld from, severely ill patients. In addition, Caplan often played the role of bioethical scold, such as disparaging plastic surgeons who performed different face-lift procedures on either side of patients' faces without the patients' consent to determine which was more effective.

Expert sources are expected to have authoritative opinions that will inspire the confidence of news consumers (Mepham, 2005). But Caplan's predominance raises the question of how a single bioethicist, no matter how distinguished, can be an expert on so many topics and whether that expertise is so superficial as to lack any real authority. Caplan views himself as a generalist but notes that bioethics is much more specialized today than when he entered the field in the 1970s.

Reporters typically use bioethicists to promote or refute a certain ethical position rather than to provide a range of ethical perspectives. This supports the assertion of bioethics scholars that reporting on bioethical issues "often seems to involve seeking out the opinion of one 'bioethics expert' and presenting it at least tacitly as representing the views of all who are in this line of work" (K. W. Goodman, 1999, p. 193). Wade acknowledged that "one bioethicist can't encompass the range of views around bioethical issues." But he differentiated between stories in which bioethics is just an aspect—when a sole bioethicist may be "sufficient to bring bioethical issues to readers' attention"—and stories in which bioethics is the main focus.

When a topic is controversial, reporters are trained to follow the journalistic norms of objectivity and balance by pairing experts representing opposing views (Boyce, 2006; Conrad, 1999; Steele, 1995). Some bioethics scholars assert that bioethicists who are used as expert sources may be reluctant to suggest other bioethicists who could provide a contrasting opinion because they believe the only correct moral position is their own (Pence, 1999). But the reporters interviewed said space and time constraints and difficulty reaching bioethicists were the principal reasons why a single bioethicist was typically used in a story.

The journalists interviewed also said they regard bioethicists as interchangeable with other types of experts. Dembner said she didn't feel the need to contrast the views of two bioethicists and that other kinds of sources could be used for balance. "I don't have time to talk to multiple bioethicists," she said. "I use one person to speak for a community." But including the views of a single bioethicist in a story may mistakenly imply moral consensus on an issue. And since bioethicists come from a range of educational, professional, and ideological backgrounds and have different religions, their viewpoints are not identical (Tuhus-Dubrow, 2006). For example, Caplan has been a strong supporter of stem cell research and a long-time critic of the American Red Cross; his position on face transplants has changed from opposition to support (Frantz, 1996; Kolata, 2005; Zitner, 1996).

Visible Scientists, Visible Bioethicists

This study offers insights into how news routines influence media coverage of scientific and medical issues with ethical implications. Just as there are "visible scientists" who play influential roles in media coverage of science (Goodell, 1977), so it appears there are "visible bioethicists" who help mold media coverage of bioethical issues. Caplan and only a few other bioethicists have become frequent expert sources in science and medical stories in six US newspapers due to their succinct quotes, accessibility, and understanding of journalistic news routines. Over time, they become news shapers by providing comment and context for stories (Conrad, 1999; Soley, 1994). The backgrounds, religions, views, and biases of this handful of bioethicists may have an indelible impact on stories in which they are quoted and on media discourse on bioethical issues overall. This is because stories tend to cascade vertically within the news hierarchy as elite newspapers and newswires take the lead in setting the media agenda (Gitlin, 1980; Nisbet et al., 2003; Rogers, Dearing, & Chang, 1991).

The reporters interviewed said a key reason why they often use bioethicists as expert sources is that they perceive bioethicists to be keen thinkers: "Thoughtful people who know an awful lot," according to Hall, formerly of the *San Francisco Chronicle*. Yet this study suggests that bioethics has been transformed from philosophy to punditry in journalism. Caplan has been extremely popular with reporters precisely because he doesn't speak in measured and scholarly tones, but in vibrant, folksy soundbites. Caplan understands news routines and is a willing co-conspirator in giving reporters what they want—a concise, dramatic, memorable quote on deadline—even if this may risk oversimplifying complex issues (Rachels, 1991).

Reporters follow news routines by using the same few trusted bioethicists over and over in a deadline-fueled search for a "dial-a-quote," as Ackerman of the *Houston Chronicle* put it. This can lead to a handful of bioethicists being asked to comment on myriad issues—in Caplan's case, on 104 topics in 188 stories—whether squarely within their scope of expertise or not. The reporters interviewed recognized that Caplan was overused as an expert source. Yet Caplan was quoted in a similar percentage of stories near the beginning and end of the 19-year study period, indicating that his prevalence as an expert source in these six newspapers had not waned.

The reporters interviewed said they believe bioethicists have a unique expertise. "Even their job title suggests that bioethicists do something special in trying to bring a moral code to science," said Deirdre Shesgreen, a former Washington correspondent for the *St. Louis Post-Dispatch*. But the reporters said quoting a single bioethicist, which occurred in 77.0 percent of the stories, was appropriate as long as their viewpoint was balanced by another type of expert. These were usually scientists, physicians, or government officials, who may have different training, experience, and expertise than bioethicists. The reporters also said they realized that bioethicists have diverse backgrounds, ideologies, and biases. But they were unconcerned about whether a story might be skewed if it contained the perspective of a single bioethicist—and especially the same bioethicist, again and again.

Caplan has been the de facto representative of the bioethics profession in the US news media for more than two decades. His accessibility, knowledge, and articulateness, combined with a genuine interest in public engagement and willingness to follow news routines, have earned him the nickname "Dr. Soundbite" (Kruvand, 2012). But concerns that Caplan's opinions, values, and biases may be conflated with those of the entire bioethics community raise questions about how the predominance of an expert source in media discourse may influence the understanding of bioethical issues among news consumers.

A Broader Spectrum

How can time-crunched journalists be encouraged to expand the range of expert sources they contact? And how can more scientists, physicians, and bioethicists be persuaded to talk with reporters? Much is at stake since neither profession can accomplish the important task of communicating with the public about science and medicine without the other. Although there are no easy answers, findings of this study provide several suggestions.

First, journalists need to acknowledge that there is an ethical element in selecting experts, and that overreliance on a limited number of expert sources can affect the objectivity and balance of science and medical stories. Searching the media organization's news archives can determine whether experts are overused by analyzing the number of expert sources, who they are, where they are from, and how often they are quoted on specific topics. If patterns of habitual use emerge, it may signal a need to find additional experts.

Second, the best time to identify and develop working relationships with new expert sources is before they're needed. That's because impending dead-lines on major stories may prompt reporters to slip into old habits and con-tact a frequently used expert who can be counted on to respond quickly. Re-porters can meet new experts at professional conferences and meetings and during visits to research labs, universities, hospitals, and other facilities. They can also seek recommendations from public relations practitioners at insti-tutions and use online services that connect reporters with sources, such as ProfNet and HARO (Help a Reporter Out).

Third, it's time for an attitude change within the scientific community. Eighty-four percent of the members of the American Association for the Ad-vancement of Science surveyed in 2015 said limited scientific knowledge of the public was a major problem for science (Pew Research Center, 2015). But who better than scientists to help increase Americans' knowledge of science? Scientists can shun talking with reporters and continue to complain about the public's scientific ignorance, or they can be part of the solution. Since the news media is a critical vehicle for communicating scientific and medical information to the public (Conrad, 1999), that means participating in media discourse.

Fourth, more scientists, physicians, and bioethicists need to overcome their reluctance to "venture out of their ivory towers" and their tendency to communicate only with their peers (Fagin, 2005). While reporters often visit hospitals, research labs, and offices of government agencies, they say it is rare to see a scientist in the newsroom (Hayes & Grossman, 2006). A desk-side briefing with a reporter is considerably less stressful than a stand-up televi-sion interview. Yet it can pay dividends in increasing the journalist's knowl-edge and in developing a working relationship based on mutual trust.

Finally, training and preparation greatly improves the likelihood of suc-cess. Some scientists, physicians, and bioethicists who avoid talking with reporters like to tell horrific anecdotes of colleagues or peers whose comments were oversimplified, taken out of context, or hyped in stories. However, me-dia training and thorough preparation can help an expert deliver his or her messages clearly and effectively while remaining in control of the interview. Media training is usually available from institutions and professional socie-ties. And as Caplan notes, the more often experts talk with reporters, the better and more efficient they become.

References

Albaek, E., Christiansen, P. M., & Togeby, L. (2003). Experts in the mass media: Researchers as sources in Danish daily newspapers, 1961–2001. *Journalism and Mass Communication Quarterly, 80*(4), 937–948.

Audit Bureau of Circulations. (2010). eCirc for newspapers. Retrieved from http://abcas3
    .accessabc.com/ecirc/newsform.asp

Boyce, N. (2002). A view from the Fourth Estate. *Hastings Center Report, 32*(3), 16–18.

Boyce, T. (2006). Journalism and expertise. *Journalism Studies, 7*(6), 889–906.

Brown, J. D., Bybee, C. R., Wearden, S. T., & Straughan, D. M. (1987). Invisible power: News-
    paper news sources and the limits of diversity. *Journalism Quarterly 64*(1), 45–54.

Burkett, W. (1986). *News reporting: Science, medicine and high technology.* Ames: Iowa State Uni-
    versity Press.

Conrad, P. (1999). Uses of expertise: Sources, quotes, and voice in the reporting of genetics in
    the news. *Public Understanding of Science, 8,* 285–302.

Conrad, P. (2001). Media images, genetics, and culture: Potential impacts of reporting scientific
    findings on bioethics. In B. Hoffmaster (Ed.), *Bioethics in social context* (pp. 90–111). Phila-
    delphia, PA: Temple University Press.

Dentzer, S. (2009). Communicating medical news—pitfalls of health care journalism. *New
    England Journal of Medicine, 360*(1), 1–3.

Dunwoody, S. (1987). Scientists, journalists, and the news. *Chemical & Engineering News, 65*(46),
    47–49.

Dunwoody, S., & Ryan, M. (1987). The credible scientific source. *Journalism Quarterly, 64,*
    21–27.

Evans, J. H. (2002). *Playing God? Human genetic engineering and the rationalization of the public
    bioethical debate.* Chicago, IL: University of Chicago Press.

Fagin, D. (2005). Science and journalism fail to connect. *Nieman Reports, 59*(4), 59–60.

Foreman, J. (1993, March 28). Cryonics draws a frozen few in New England. *The Boston Globe,*
    p. 1A.

Frantz, D. (1996, May 30). Blood bank politics—a special report; Elizabeth Dole and Red Cross:
    2 powers at work. *The New York Times,* p. A1.

Friedman, S. M., Dunwoody, S., & Rogers, C. L. (Eds.). (1986). *Scientists and journalists: Report-
    ing science as news.* New York, NY: Free Press.

Gans, H. J. (1979). *Deciding what's news.* New York, NY: Pantheon Books.

Gitlin, T. (1980). *The whole world is watching: Mass media in the making and unmaking of the
    New Left.* Berkeley: University of California Press.

Goodell, R. (1977). *The visible scientists.* Boston: Little, Brown.

Goodell, R. (1986). How to kill controversy: The case of recombinant DNA. In S. M. Friedman,
    S. Dunwoody, & C. L. Rogers (Eds.), *Scientists and journalists: Reporting science as news*
    (pp. 170–181). New York, NY: Free Press.

Goodman, E. (1997, February 6). Expertise deflation. *The Boston Globe,* p. A19.

Goodman, K. W. (1999). Philosophy as news: Bioethics, journalism and public policy. *Journal of
    Medicine and Philosophy, 24*(2), 181–200.

Gregory, J., & Miller, S. (1998). Working with the media. In A. Wilson (Ed.), *Handbook of Science
    Communication* (pp. 79–89). Bristol, UK: Institute of Physics.

Guthmann, E. (2006, May 31). Stanford go-getter has become go-to guy for those facing quan-
    daries over new medical technologies. *The San Francisco Chronicle,* p. E1.

Hartz, J., & Chappell, R. (1997). Worlds apart: How the distance between science and jour-
    nalism threatens America's future. First Amendment Center, Vanderbilt University. Re-
    trieved from http://www.firstamendmentcenter.org/madison/wp-content/uploads/2011/
    03/worldsapart.pdf

Hayes, R., & Grossman, D. (2006). *A scientist's guide to talking with the media*. New Brunswick, NJ: Rutgers University Press.

Ipaktchian, S. (2011). Bioethics at midlife. *Stanford Medicine, 28*(1), 8–13.

Kolata, G. (2005, December 16). Clone scandal: "A tragic turn" for science. *The New York Times*, p. A6.

Kovach, B., & Rosenstiel, T. (2001). *The elements of journalism: What newspeople should know and the public should expect*. New York, NY: Crown.

Kruvand, M. (2009). Bioethicists as expert sources in science and medical reporting. *Newspaper Research Journal, 30*(3), 26–41.

Kruvand, M. (2012). "Dr. Soundbite": The making of an expert source in science and medical stories. *Science Communication, 34*(5), 566–591.

Levine, C. (2007). Analyzing Pandora's box: The history of bioethics. In L. A. Eckenwiler and F. G. Cohn (Eds.), *The ethics of bioethics: Mapping the moral landscape* (pp. 3–23). Baltimore: Johns Hopkins University Press.

Limoges, C. (1993). Expert knowledge and decision making in controversy contexts. *Public Understanding of Science, 2*, 417–426.

Maier, S., & Kasoma, T. (2005, May). Information as good as its source—an examination of source diversity and accuracy at nine daily U.S. newspapers. Presentation to the International Communication Association, New York, NY.

Martin, S. E. (1991). Using expert sources in breaking science stories: A comparison of magazine types. *Journalism Quarterly, 68*(1/2), 179–190.

Meier, B., & Thomas, K. (2015, May 7). Eager to opine on the toughest calls in medical ethics. *The New York Times*, p. B4.

Mepham, B. (2005). *Bioethics: An introduction for the biosciences*. Oxford, UK: Oxford University Press.

Nelkin, D. (1995). *Selling science: How the press covers science and technology* (revised ed.). New York, NY: Freeman.

Nelkin, D. (2001). Beyond risk: Reporting about genetics in the post-Asilomar press. *Perspectives in Biology and Medicine, 44*(2), 199–207.

Nisbet, M. C., Brossard, D., & Kroepsch, A. (2003). Framing science: The stem cell controversy in an age of press/politics. *Press/Politics, 8*(2), 36–70.

Nisbet, M. C., & Lewenstein, B. V. (2002). Biotechnology and the American media: The policy process and the elite press, 1970 to 1999. *Science Communication, 23*(4), 359–391.

Pence, G. E. (1999). The bioethicist and the media. *Princeton Journal of Bioethics, 2*(1), 47–52.

Petit, C. (2007). Science writing: The changing landscape. *Symmetry, 4*(3), 34.

Pew Research Center. (2015, January 29). Public and scientists' views on science and society. Retrieved from http://www.pewinternet.org/2015/01/29/public-and-scientists-views-on-science-and-society/

Pinheiro, D. (2015). How can we redesign newsrooms to be digital-first without losing their journalistic soul? John S. Knight Journalism Fellowships at Stanford University. Retrieved from http://jsk.stanford.edu/journalism-challenges/2015/how-can-we-redesign-newsrooms-to-be-digital-first-without-losing-their-journalistic-soul/

Rachels, J. (1991). When philosophers shoot from the hip. *Bioethics, 5*(1), 67–71.

Reese, S. D., & Danielian, L. H. (1994). The structure of news sources on television: A network analysis of "CBS News," "Nightline," "MacNeil/Lehrer" and "This Week with David Brinkley." *Journal of Communication, 44*(2), 84–107.

Reich, Z. (2006). The process model of news initiative: Sources lead first, reporters thereafter. *Journalism Studies, 7*(4), 497–514.

Rogers, E., Dearing, J., & Chang, S. (1991). AIDS in the 1980s: The agenda-setting process of a public issue. *Journalism Monographs*, 126.

Rosenfeld, A. (1999). The journalist's role in bioethics. *Journal of Medicine and Philosophy 24*(2), 108–129.

Rubin, D. M., & Hendy, V. (1977). Swine influenza and the news media. *Annals of Internal Medicine, 87*, 769–774.

Russell, C. (2006). Covering controversial science: Improving reporting on science and public policy. Joan Shorenstein Center on the Press, Politics and Public Policy. John F. Kennedy School of Government, Harvard University. Retrieved from http://www.hks.harvard.edu/presspol/publications/papers/working_papers/2006_04_russell.pdf

Shepherd, R. G. (1981). Selectivity of sources: Reporting the marijuana controversy. *Journal of Communication, 31*, 129–137.

Shoemaker, P. J., & Reese, S. D. (1996). *Mediating the message: Theories of influences on mass media content.* White Plains, NY: Longman.

Simonson, P. (2002). Bioethics and the rituals of the media. *Hastings Center Report, 32*(1), 32–39.

Soley, L. C. (1994). Pundits in print: "Experts" and their use in newspaper stories. *Newspaper Research Journal, 15*(2), 65–75.

Steele, J. E. (1995). Experts and the operational bias of television news: The case of the Persian Gulf War. *Journalism & Mass Communication Quarterly, 72*(4), 799–815.

Stocking, S. H. (1985). Effects of public relations efforts on media visibility of organizations. *Journalism Quarterly, 62*(2), 358–366, 450.

Stolberg, S. G. (2002, May 5). It's alive! It's alive! *The New York Times*, p. D16.

Sumpter, R. S., & Garner, J. T. (2007). Telling the Columbia story: Source selection in news accounts of a shuttle accident. *Science Communication, 28*(4), 455–475.

Tanner, A. H. (2004). Agenda building, source selection, and health news at local television stations: A nationwide survey of local television health reporters. *Science Communication, 25*(4), 350–363.

Tenore, M. J. (2009, December 14). Angier: Newspaper science reporting is "basically going out of existence." Poynter Online. Retrieved from http://www.poynter.org/column.asp?id=101&aid=174722&view=print

Toner, R. (2001, January 21). The abortion debate, stuck in time. *The New York Times*, p. D1.

Tuchman, G. (1978). *Making news: A study in the construction of reality.* New York: Free Press.

Tuhus-Dubrow, R. (2006, July 25). Doctors without borders: Bioethics matures into a formal academic field and faces an identity crisis. *The Village Voice.* Retrieved from http://www.villagevoice.com/arts/doctors-with-borders-7143680

Van Dijk, T. A. (2004). About a year before the breakdown I was having symptoms. In Clive Seale (Ed.), *Health and the media* (pp. 160–175). Oxford, UK: Blackwell.

Zitner, A. (1996, March 24). Blood feud: New rules in battle for donors. *The Boston Globe*, p. B1.

8

The Ethics and Boundaries of Industry Environmental Campaigns

BARBARA MILLER GAITHER
AND JANAS SINCLAIR

In 2009, ExxonMobil announced plans to invest $600 million to explore the renewable energy possibilities of algae-based biofuels in a partnership with biotech company Synthetic Genomics. The oil giant's collaboration with Synthetic Genomics was slated to last five to six years and entail significant research into algae as a source of oil that could be converted to conventional transportation fuels, including gasoline and diesel. Ultimately, after just four years and $100 million invested, the company announced that it would refocus its research with Synthetic Genomics after the work failed to produce economically viable results (Herndon, 2013).

At the same time ExxonMobil began its initial foray into biofuels research, it also launched a widespread advertising campaign to promote the company's investments in alternative sources of energy, along with the company's environmental advocacy. One of the television ads for "Advanced Biofuels" featured a scientist, Joe, serving as an organizational representative for the company. In the ad, Joe explains that, "in using algae to form biofuels, we're not competing with the food supply, and they absorb CO_2, so they help solve the greenhouse problem as well."

This advertisement was aired extensively in the United States, particularly on American primetime news and sports channels, but was banned by the UK Advertising Standards Authority (ASA) for overstating the climate change mitigation potential from algae-based technology. The initial complainant objected to the fact that while the ad implied that the new technology would reduce CO_2 levels, the carbon dioxide absorbed by algae would actually be released back into the atmosphere when it was burned as fuel. Although Exxon-Mobil appealed the ruling as "inconsistent with expert opinion on the role of advanced biofuels in addressing growth in greenhouse gas emissions" (Eur-active, 2011), UK's ASA upheld the ban on the ad stating that it "overstated the technology's total environmental impact and was therefore misleading" (ASA, 2011).

The case of ExxonMobil's "Advanced Biofuels" ad exemplifies the chal-lenges and concerns, both from an environmental and a corporate perspective, of what has come to be known as "greenwashing." The term "greenwashing" is often used to describe the tendency of corporate advocacy campaigns to exaggerate the extent and effects of corporate environmental initiatives, and in many cases, promote products and production processes that are neither environmentally friendly nor sustainable (Plec & Pettenger, 2012). While en-vironmental advocates seek to call attention to the implications and conse-quences of corporate greenwashing, businesses that are genuinely committed to pro-social and environmental causes may find it challenging to commu-nicate an environmental message without facing heightened criticism and backlash. All the while, audiences are likely finding it increasingly difficult to distinguish between "those companies genuinely dedicated to making a difference and those using a green curtain to conceal dark motives" (Green-peace, 2013).

The following sections highlight industry environmental campaigns that exemplify marketplace advocacy, a form of issue advocacy that is especially susceptible to charges of greenwashing—and often justifiably so. The case of ExxonMobil's "Advanced Biofuels" campaign and other corporate- and industry trade group-sponsored examples illustrate the ethical challenges as-sociated with marketplace advocacy. Finally, the boundaries for appropriate and ethical industry environmental campaigns are identified with guidelines for responsible promotion of industry environmental initiatives. Recogni-tion of these guidelines for responsible industry advocacy may ultimately allow corporations genuinely committed to environmental care to move be-yond the "greenwashing" critique and be recognized for legitimate pro-social and environmental efforts.

There are three key issues of particular concern for industry environmen-tal advocacy, issues often circumnavigated by many environmental advocacy campaigns. These include transparency of the advertising sponsor, accuracy in describing industry activities, and responsible use of values appeals. While industry campaigns frequently fall short in providing transparency and accuracy, they do typically focus on values, which empirical research suggests may have detrimental consequences for environmental initiatives (Miller & Sinclair, 2014), and therefore require particular attention for ethical usage.

Industry Environmental Campaigns

At the broadest level, industry environmental campaigns sponsored by businesses and industry trade groups can be classified as an example of corporate marketing. While traditional marketing focuses on fulfilling customer needs through products or services, corporate marketing has a broader focus on stakeholders and their ongoing relationship with the organization (Balmer, Powell, & Greyser, 2011). In traditional marketing, ads are designed to sell a product or service, but in corporate marketing, ads are designed to communicate about the organization to a variety of stakeholders regarding issues beyond the immediate goal of stimulating sales.

The genre of "green marketing" can be segmented along similar lines—as either promoting the relationship between a product or service and the environment to stimulate sales (e.g., Timberland's "Earthkeepers" campaign showcasing eco-friendly apparel), or presenting a corporate image of environmental responsibility to stimulate goodwill (Banerjee, Gulas, & Iyer, 1995). Industry environmental campaigns represent the latter category, as messages emphasize corporate (or industry) image rather than stimulating demand for a particular product or service. ExxonMobil's "Advanced Biofuels" advertisement, for example, was not selling a product, nor was it an effort to promote algae-based fuels per se. As the ad ran prominently on mainstream television, the vast majority of viewers had little knowledge or expertise in the topic being discussed and were not being called upon to purchase biofuel. Rather, the advertisement was selling an image of ExxonMobil as an environmental proponent.

While industry environmental campaigns are frequently labeled as Corporate Social Responsibility (CSR) in academic literature, most of these campaigns, in fact, represent a unique genre of corporate issue advocacy, known as marketplace advocacy (Miller & Lellis, 2015; Miller & Sinclair, 2009a, 2009b). CSR involves corporations' "voluntary consideration of stakeholder concerns both within and outside its business operations" (Homburg, Stier, & Bornemann, 2013, p. 54), so CSR campaigns address pro-social initiatives in a range of areas. For example, CSR communication includes Target Corporation's "It Comes from the Heart" ad about Target employees volunteering to distribute food to fight childhood hunger and Yoplait Yogurt's "Save Lids to Save Lives" campaign about their initiative to raise money for the Susan G. Komen organization for breast cancer.

Both CSR and marketplace advocacy are intended to communicate a positive brand image and portray the company as responsive to the needs of society while also serving the needs of the company. What makes marketplace advocacy distinct from CSR, however, is an emphasis on protecting the organization's position in the marketplace. In marketplace advocacy, the corporate good works featured in the message are directly tied to the organization's own source of profit and protecting that profit. This includes generating public support to avoid calls for potential government regulation of the business or industry. Marketplace advocacy fosters the impression that voluntary corporate efforts make regulation unnecessary (Sinclair & Miller, 2012). For example, campaigns have been used by American Electric Power to reduce regulations on coal mining, by Mobil Oil to prevent legislative passage of an excess profits tax directed at oil companies, and by the Chrysler Corporation to slow the implementation of automotive pollution controls (Cutler & Muehling, 1989).

Unlike other forms of pro-social CSR initiatives, industry environmental campaigns, or more accurately, marketplace advocacy campaigns, focus squarely on the company's own business, whether its current products or services, future offerings in development, or the processes used to create these products and services. ExxonMobil's "Algae" ad, for example, focuses on their algae biofuel research and its potential to reduce greenhouse emissions. While this ad may portray ExxonMobil as responsive to society's needs, it does so to protect the company's position in the marketplace by assuaging current or potential stakeholder concerns about its environmental impact.

Could a company like ExxonMobil, with business practices that have wide-ranging environmental implications, engage in an environmental campaign that is CSR, and not marketplace advocacy? Theoretically, the answer is "yes," as long as the environmental initiative is not related to the company's business activities. Campaigns supporting species conservation or recycling efforts, for example, could potentially be CSR for an energy company. Specifically in the case of ExxonMobil, the company's "Algae" ad is qualitatively different from ExxonMobil's CSR sponsorship of "Think It Up," a multi-year education initiative to generate excitement for education and raise awareness of national education concerns. While the "Think It Up" CSR campaign could also be expected to accrue benefits to the organization, it does not focus on protecting or promoting the company's own source of profit.

In sum, industry environmental campaigns can be defined as marketplace advocacy when they are (a) sponsored by a corporation (or an industry trade organization representing a group of corporations) and (b) communicate information about corporate environmental initiatives or make environmental claims about corporate practices, with the underlying goals of (c) building positive attitudes toward the sponsor, maintaining a climate supportive of the sponsor and its business activities, and reducing potential for future government intervention in corporate activities. Also important to note, given corporate financial resources, these campaigns often constitute a significant share of voice in the arena of environmental messages.

Marketplace Advocacy and Greenwashing

While the business of green marketing in general is thriving—with sales topping $56 billion in 2006 (Marty, 2007)—there are also concerns about questionable product claims. And while green product advertising has come under fire, industry environmental claims, as opposed to product ads, are particularly subject to charges of greenwashing when the advocated corporate image doesn't match its environmental performance (Corbett, 2006). According to Greenpeace (2013), for some businesses, efforts to promote an ecological conscience may be "little more than a convenient slogan. . . . At best, such statements stretch the truth; at worst, they help conceal corporate behavior that is environmentally harmful by any standard."

In ExxonMobil's "Advanced Biofuels" campaign, the company shifts from energy supplier to environmental advocate. On the surface, this advertising effort to promote itself as "green" is neither oppressive nor manipulative; however, the company has consistently resisted shifts away from petroleum-based fundamentals and has publicly questioned the science and significance of climate change (Plec & Pettinger, 2012). A 2012 press release from the Heartland Institute—an organization that received funding from ExxonMobil and other fossil fuel interests—stated, "The claim that there is a 'scientific consensus' that global warming is both man-made and a serious problem is untrue" (Union of Concerned Scientists, 2014). Meanwhile, in 2015, the CEO of the company mocked renewable energy in a shareholder speech, stating his firm hadn't invested in renewable energy "because we choose not to lose money on purpose" (Lerner, 2015). In other words, ExxonMobil's mainstream advertising is designed to build positive attitudes toward the corporate sponsor and maintain support for business activities through environmental claims, regardless of actual corporate practices.

Industry environmental campaigns constituting marketplace advocacy may be particularly susceptible to greenwashing charges given the common goal of reducing the potential for future government intervention in corporate activities (e.g., Cutler & Muehling, 1989; Sethi, 1977). These campaigns can help accomplish this goal by downplaying the industry's negative impacts and promoting the positive message that corporate initiatives can address environmental concerns. In fact, ExxonMobil's "Advanced Biofuels" advertisement, which seeks to persuade consumers that it is a responsible caretaker of the environment (Plec & Pettinger, 2012), is demonstrative of messaging used throughout the energy sector. Schlichting (2013) found that between 1990 and 2010, a common industry framing strategy used by the US fossil fuel and electric utility industries was to promote uncertainties regarding climate science as well as the socioeconomic consequences of regulatory climate policies and treaties that might negatively impact the industries.

Even when corporate environmental efforts are voluntary—which is not always the case, as the environmental change may have been initiated under threat of governmental prosecution—the millions of dollars spent on the marketplace advocacy campaign to tout environmental achievements can overshadow the cost of minor environmental improvements (Sethi, 1977). Since 2002, ExxonMobil invested a total of $188 million into alternative energy, compared to $250 million spent on US advertising from 2010 to 2012 alone. Meanwhile, in 2012 the company made approximately $45 billion in profit (Juhasz, 2013).

Despite these examples, however, some businesses are genuinely committed to environmental and social responsibility. The challenge for corporations in communicating environmental messages is avoiding the greenwashing label without clear guidelines as to what constitutes ethical corporate environmental communication. While lists of "greenwashing sins" for product-based environmental claims are available from business and marketing firms (e.g., TerraChoice), the following guidelines are specifically relevant to industry environmental claims, particularly those representing marketplace advocacy.

Ethical Marketplace Advocacy: Is It Possible?

While responsible advertising and public relations tactics are certainly possible, industry environmental campaigns require extra vigilance on the part of advertising and public relations practitioners to ensure responsible communication. While all types of corporate marketing require ethical consideration, this is particularly true for marketplace advocacy for a number of reasons. Although marketplace advocacy efforts resemble CSR communication in portraying a company as serving the needs of society, they function to protect the company's own source of profit, and therefore serve the distinct goal of marketplace advocacy. The potential for this intention (i.e., protecting the company's own position in the marketplace) to be portrayed primarily as the intention to serve social interests indicates marketplace advocacy campaigns warrant particular ethical consideration. As the researcher L'Etang (1996) describes public relations ethics, "there is clearly something wrong about claiming moral capital while at the same time being driven largely by self-interest" (p. 91). Marketplace advocacy campaigns seek to build a virtuous brand image, but are largely intended to promote a policy agenda. These underlying policy objectives underscore the need for ethical guidelines for this category of communication.

Moreover, the current environmental crisis itself calls for careful ethical consideration of all corporate environmental messages. Current scientific consensus indicates the planet is experiencing the effects of climate change and that these changes signal the beginning of an impending environmental crisis. The White House's Third National Climate Assessment (May, 2014) and the United Nation's Intergovernmental Panel on Climate Change report (April, 2014) call for dramatic changes in energy usage to achieve significant reduction in carbon emissions, and the next 15 years have been identified as critical for enacting these carbon-policy changes (Global Commission on the Economy and Climate, 2014). Whether or not these changes are enacted could very well hinge on public awareness and pressure exerted on policy makers. The stakes for global health are high, and public saliency of the need for carbon policy change may have a significant impact on the future of climate change. It seems appropriate that corporate claims that may influence perceptions about the need to adopt changes should receive particular ethical scrutiny.

In the next section, boundaries for responsible industry environmental campaigns are presented. The following discussion is based on a synthesis of existing ethical recommendations from the literature as they apply to the context of industry environmental campaigns. Criteria that are especially relevant for these campaigns are presented rather than an exhaustive description of ethical advocacy. And while industry environmental campaigns may include a number of communication strategies, these guidelines focus primarily on advertising, because it represents a key strategy for reaching national, lay audiences with little or no expertise in the advocated topic. Three key criteria are identified: (1) transparency of the message source, (2) accuracy in describing industry activities, and (3) responsible use of values appeals.

Transparency of the Message Source: Visibility

Transparency has been identified as an ideal in the communications profession. Professional organizations in public relations and communication (including the Public Relations Society of America, the International Public Relations Association, the International Association of Business Communicators, and the American Marketing Association) underscore the need for truthful and full disclosure of information in three areas: between client and organization, between employee and employer, and between organization and community. This last area, communication between the organization and the community, is the domain of concern for ethical industry environmental campaigns. The roots of these professional guidelines run deep, drawing on the basic tenet forwarded by philosopher Immanuel Kant that people should be treated as "ends" rather than as only a means to another's goal. Academic literature in public relations has identified transparency of the message source as a criterion for ethical public relations advocacy. Edgett (2009), for example, calls this the criterion of visibility, or "clear identification of all communications on behalf of the client or organization as originating from that source" (p. 19). If an organization fails to reveal its identity to gain an advantage, or orchestrates a campaign to present its views as coming from some other source, then it fails to act responsibly to audience members. Failing to satisfy the requirement of visibility could even be considered harmful to democracy, because it implies a lack of fairness in public debate (Edgett, 2009).

While identification of the message source is not usually an issue in advertising where brand recognition is a key outcome for advertisers, corporate environmental campaigns are often sponsored by industry associations and have been criticized not only for lack of visibility, but in some instances, for intentionally obscuring the message source. The terms "front group" and "stealth front group" have been used to describe lack of transparency among industry organizations and identified as an element of deception in greenwashing (Laufer, 2003). For example, in the past, ExxonMobil funded the National Wetlands Coalition, an organization made up of companies, including land developers, working to *oppose* federal efforts to restrict wetlands' development. This creation of a seemingly independent third-party organization with a name that hides the true interests of the organization can be considered unethical at face.

Many industry environmental campaigns, however, are often not disguised behind names that suggest objectivity. Rather, industry associations that sponsor media campaigns are typically advocating for protecting the industry represented by the trade association name. The message source is fairly transparent, for example, in the "Energy Tomorrow" campaign by the American Petroleum Institute that advocates for the oil and natural gas industries. Nevertheless, in order to be considered ethical and avoid potential "greenwashing" charges, the burden of identification should rest with the campaign sponsor. Even in situations where the message source is fairly transparent, membership within the groups may not be. For complete transparency, industry groups should identify their membership fully through media options with unlimited time and space, such as websites. In media options that are more constrained with regards to time and space, advertising should identify the membership as clearly as possible. For example, the American Petroleum Institute could satisfy the criterion for visibility in a TV ad by indicating the message is sponsored by companies that produce oil and natural gas.

Accuracy in Describing Industry Activities: Veracity

Perhaps the most common criticism of industry environmental campaigns is that they are guilty of "greenwashing" by presenting exaggerated or unwarranted claims of sustainability or positive environmental impact (Dahl, 2010; Laufer, 2003). Thus, accuracy in describing industry activities is the second criterion for responsible industry environmental campaigns. The term "veracity" has been used to describe this criterion in the literature on ethical public relations advocacy and is defined as full truthfulness in all matters (Edgett, 2009). The criterion of veracity has also been discussed specifically in the context of communicating corporate identity, with the imperative that corporate identity ("what we really are") should be congruent with communication ("what we say we are") to avoid deception (Fukukawa, Balmer, & Gray, 2007).

According to Sher (2011), a marketing tactic is deceptive if it is intended to "bring about consumer misconception by providing what the marketer believes is false evidence, omitting key evidence, or misrepresenting what the evidence means" (p. 104). Deceptive tactics are considered manipulative and immoral based on the fundamental tenet that people should be treated as an end rather than a means to a goal. Deception is considered moral only when there is some type of redemptive moral consideration, such as all truthful possibilities having being ruled out (Edgett, 2009). Communication practitioners must often be selective in the information presented in a particular campaign message due to limitations of media time and space, thus the concept of "substantial completeness" has been used to define an acceptable amount of information disclosure in mass communication advocacy situations (Martinson, 1996). A related concern is "spin," or selecting those facts that present an overly positive light, which can be considered a form of deception if it is intended to bring about consumer misconception (Sher, 2011).

VERACITY IN ENVIRONMENTAL IMPACT CLAIMS

For responsible industry environmental campaigns, advertisers must accurately present claims of sustainability or positive environmental impact.

As mentioned, the ExxonMobil ad about the company's research on using algae as a biofuel was banned by the UK's ASA for overstating the potential of the technology to reduce CO_2 levels. In the ad, Joe the scientist explained that the use of biofuels from algae helps "solve the greenhouse problem" as the algae does not compete with the food supply and also absorbs CO_2. The ASA determined this ad overstated the potential environmental impact of algae as a biofuel and was therefore misleading. Although it is true that algae biofuel would not release new CO_2 into the atmosphere, the CO_2 absorbed by growing algae would be released when it is burned as fuel, and therefore the process would be a break-even proposition, as opposed to yielding a net outcome of absorption (ASA, 2011). While ethical standards are conceptually distinct from regulatory standards, this example illustrates the accuracy in language required for truthful environmental claims.

VERACITY IN PORTRAYING ACTIVITIES AS VOLUNTARY VS. COMPULSORY

To achieve veracity, industry environmental campaigns must accurately present claims of sustainability or positive environmental impact. An additional concern related to the accuracy of these campaigns is the tendency to portray environmental initiatives as voluntary while, in fact, the industry's response may have been government mandated or initiated under threat of government prosecution. Such campaigns attempt to downplay an industry's adverse effects on the environment by exaggerating the efforts of industries to control pollution—and in some cases publicize adverse effects to the economy that may result from various regulatory efforts (Sinclair & Miller, 2012; Sethi, 1977).

One example is BP's campaign following the Deepwater Horizon oil catastrophe in the Gulf of Mexico in 2010. Ads discussed clean-up activities and highlighted the amount of money spent on clean-up and claims, but did not mention that these actions were required by a legally binding settlement. Touting environmental improvements or initiatives that have been mandated by the government or a legally binding settlement may be considered deceptive. On the other hand, responsible industry environmental campaigns can discuss these improvements or initiatives, but rather than praising them as if they were self-imposed, the criterion of veracity calls for identifying the laws with which the company or industry complies. This could be a particular settlement, or a statement of compliance with industry environmental regulations, such as the Energy Policy Act, the Clean Water Act, and/or the Energy Conservation Act. This disclosure would satisfy the criterion of veracity concerning voluntariness, and might, in fact, provide the company with an opportunity to showcase all of the environmental regulations with which they comply—as well as highlight any instances in which they exceed those regulatory standards. Such disclosure provides an opportunity for ethical advocacy that would communicate potentially useful information about industry accountability with audience members, a strategy that has been shown to generate industry-favorable attitudes among audiences (Miller & Sinclair, 2009a, 2009b; Sinclair & Miller, 2010).

Appropriate Use of Values Appeals: Sensitivity

The third criterion for responsible industry environmental campaigns is sensitivity to social responsibility. The criterion of sensitivity requires the goals of the organization to be balanced with the needs and concerns of society (Edgett, 2009; Balmer, Powell, & Greyser, 2011). While the organization's goals are, by definition, a priority in marketplace advocacy, they must be balanced with consideration of, or sensitivity to, the greater social welfare. Sensitivity to the responsible use of values appeals is a key issue for industry environmental campaigns. Information about corporate activities is relatively limited in industry environmental campaigns; instead, the main focus typically involves associating corporate activities with commonly held social values. The use of values-based messages in industry environmental campaigns, combined with empirical findings about their effects, requires consideration of the best interests of society.

Message strategy in marketplace advocacy often praises societal values, condemns oppositional values, discusses philanthropic efforts, or associates an organization's products with worthwhile societal goals. In the 1970s, for example, Phillips Petroleum engaged in a campaign promoting the company's contribution to the public good, including the development of a blood filter for kidney patients and a fuel additive that helped make a helicopter rescue possible from a snowy mountain (Bostdorff & Vibbert, 1994). More recently, ads for the American Plastics Council's "Plastics Make It Possible" campaign similarly highlighted how the plastics industry made societal contributions, emphasizing the "everyday miracles" of plastics in child and food safety products. A common theme among many coal-related advocacy campaigns, meanwhile, involves an appeal to regional and national pride. The "Friends of Coal" campaign by the West Virginia Coal Association emphasizes "hard-working coal miners," as well as the industry's heritage and role in powering a growing nation (West Virginia Coal Association, 2009). On a national level, the "America's Power" campaign by the American Coalition for Clean Coal Electricity appeals to US nationalism, encouraging website visitors to join "America's Power Army" and promoting "clean coal" as America's most abundant energy source (American Coalition for Clean Coal Electricity, 2010).

Among the most common values-based communication strategies in industry environmental campaigns are the values of optimism, determination, and ingenuity as tools to overcome seemingly insurmountable challenges—with the implication that no drastic changes are needed in current business practices or consumption patterns. Returning to the ExxonMobil "Advanced Biofuels" example, Joe the scientist explains that the algae is "very beautiful," and describes their "blue or red, golden, green" colors. His appreciation for nature and his optimism about our energy future are apparent. Thus, the ExxonMobil ad, like many marketplace advocacy campaigns, is not selling an energy product, but rather the optimism and integrity of the sponsoring company (Plec & Pettenger, 2012), as well as the value of ingenuity to overcome challenges through science.

An ad from GE's "Ecomagination" campaign similarly references these values through the classic story, The Little Engine that Could. The print ad reads, "Can technology and the environment peacefully coexist? Ecomagination answers yes with the Evolution Series locomotive. . . . This is the little engine that could. And will." From a pro-environmental perspective, this type of message seems to embody the "fantasy of technical fixes" and the "siren call of denial" that some have argued characterizes public opinion on the environment (Shellenberger & Nordhaus, 2004, p. 5). The ethical concern is that corporate environmental campaigns may encourage the public to ignore current problems, or the need for change, with the exaggerated promise that future scientific innovation will solve environmental problems. While past scientific innovation has allowed for tremendous progress in addressing many human problems, there is no guarantee for the rate of scientific discovery or the specific problems that will be addressed in the future.

INDUSTRY ENVIRONMENTAL CAMPAIGNS AND ENVIRONMENTAL CONCERN

It seems the desire to overlook environmental problems—and to believe they can be solved by science and industry—may be reinforced by these campaigns. In fact, empirical findings based on a US national Web-based survey indicate environmental marketplace advocacy campaigns by both corporations and an industry trade group—including GE's "Ecomagination" campaign, DuPont's "Open Science" campaign, and the American Coalition for Clean Coal Electricity's "America's Power" campaign—were successful in generating message acceptance and, further, that audience members' environmental concern was *positively* correlated with both message acceptance and their attitudes toward the advertiser's environmental impact (Miller & Sinclair, 2014). Follow-up analyses revealed that environmental concern was positively correlated with all of the relationships investigated (including perceptions of the industry's commitment to society and the environment, intent of the ad message, participants' motives to share the values of the advertiser, and attitudes toward the advertiser and the advertiser's environmental impact).

These findings have significant ethical implications, because they provide evidence that industry environmental campaigns can lead the public to set aside their environmental concerns under the assumption that they are already being addressed. For example, one of the TV ads used in this study was a message from the American Coalition for Clean Coal Electricity (ACCCE), an industry group made up of companies involved in producing electricity from coal, including coal producers, utility companies, and railroads. The ACCCE ad stated "technology born from American ingenuity can achieve amazing things" and goes on to present benefits of clean coal technology in terms of "lower emissions, capture and storage of CO_2." Among participants who viewed this ad, higher levels of environmental concern led to more positive attitudes toward the advertiser's environmental impact rather than increased skepticism, or concern.

SENSITIVITY TO SOCIETAL NEEDS

Based on their potential harm to long-run social welfare in impeding action to address the environmental crisis, sensitivity to social responsibility is therefore the third criterion for responsible industry environmental campaigns. While the organization's goals are a priority in marketplace advocacy, the criterion of sensitivity requires these goals to be balanced with the needs and concerns of society, including consideration of, or sensitivity to, social welfare (Edgett, 2009; Balmer, Powell, & Greyser, 2011).

Responsible advertising has been defined as serving the interest of the advertiser, while also not harming any stakeholder. Some definitions of responsible advertising go even further, requiring advertisers to encourage behaviors that trustworthy evidence indicates are consistent with long-run social welfare and to discourage behaviors that are not (Hyman, 2009). Communication professionals should carefully consider the degree to which corporate environmental messages encourage behavior that is inconsistent with long-run social welfare. For the ExxonMobil ad on the company's research on using algae as a biofuel, sensitivity requires practitioners to consider the potential impact of the message that algae helps "solve the greenhouse problem." While this message may accrue benefits for ExxonMobil, is there potential harm to society in suggesting that this technology is close to "solving" the environmental crisis?

The potential negative impact of corporate environmental messages lies in persuading the public that environmental problems have been addressed and that increased regulation, for example, is unnecessary. In other words, environmental marketing claims can "impede finding real solutions to identified problems by causing consumers to set aside their environment concerns making the assumption that these concerns had been addressed" (Davis, 1992). If the result is inaction, including failure to adopt changes in policy related to carbon emissions, then environmental consequences can be expected to have a negative effect on the social well-being of current and future generations.

SENSITIVITY TO BALANCING VALUES AND INFORMATION

Industry environmental campaigns are based strongly on an appeal to values, and while values appeals are not irresponsible per se, they must be used in conjunction with accurate information about corporate or industry activities. While critics have debated the ethicality of marketing tactics that attempt to influence consumer decision making in ways other than providing straightforward, accurate information, scholars in marketing, advertising, and public relations generally agree that non-cognitive, or "image," appeals can be used responsibly (Sher, 2011; Hyman, 2009). In industry environmental campaigns, the appeal to the value of scientific ingenuity and its problem-solving potential—while rather emotionally presented—is not harmful to society in itself as long as it is not overstated. Messages about scientific problem solving can be considered pro-social to the degree that they are realistically inspirational and promote scientific inquiry. Technological innovation is certainly a key component to addressing the environmental crisis, even if it should not be considered as a cure-all. Values appeals may be used responsibly in industry environmental campaigns when they are balanced with substantive information and the intent is to serve societal needs as well as the needs of the organization.

A one-minute TV commercial or a display ad in an airport can communicate only so much information about environmental initiatives such as ExxonMobil's algae research program. At the same time, however, this is not an excuse to communicate only through values appeals. Relying on values appeals—without providing concrete information about the message sponsor and actual corporate activities—could easily lead to charges of greenwashing. According to the consulting group Terrachoice Environmental Marketing, the "sin of vagueness" is one of the ways that companies commit greenwashing (Terrachoice, 2009). Image, in the absence of information, is a strategic way to be vague.

From an ethical perspective, even more significant is the purpose of the values appeal and whether that message serves the interest of the audience members. Values appeals should follow this basic rule for ethical persuasion: "The practitioner attempting to truthfully persuade should genuinely believe that he or she is assisting the receiver in attaining that which the receiver already implicitly seeks and is in the receiver's interest" (Martinson, 1996, p. 44). The next section describes the need for sensitivity to receiver, or audience member's, goals.

SENSITIVITY TO AUDIENCE GOALS

In the case of industry environmental campaigns, values appeals are clearly unethical if they are designed to subvert audience members' environmental concern and encourage them to act against their personal and societal interests—or *not* act to pursue those interests. Attempting to undermine people's decision-making processes has been specifically identified as an unethical marketing tactic, and it becomes a form of manipulation when the attempt involves influencing others' behavior by altering their goals (Rudinow, 1978). Manipulation occurs either by some type of deception, as discussed in the criteria of visibility and veracity, or by playing on a weakness in the consumer's normal decision-making process. Marketing tactics are manipulative if they are intended to undermine the audience member's level of rationality in a particular context or reduce the amount of helpful information available (Sher, 2011). While the purpose, or intent, has been identified as key in determining whether a communication tactic is manipulative, an organization can still be considered morally blameworthy for generating a change in stakeholders' beliefs, desires, or preferences that is detrimental to society (Sher, 2011). While stakeholders themselves undoubtedly have some personal responsibility in their reaction to advocacy messages, this does not relieve the organization of the need to be sensitive to its social responsibility.

On the basis of the above, responsible industry environmental campaigns should not use values appeals with the intent of playing on audience members' vulnerabilities or to undermine audience members' abilities to make good decisions that would be consistent with their environmental concern. In the case of ExxonMobil's ad about algae, the optimism of an individual scientist talking about the beautiful colors of algae should not be used if the intent is to undermine the concern people may have for the environment and the environmental crisis. While image- or values-based appeals are not in and of themselves unethical, providing accurate information to the public would clearly aid the responsible communicator in advocating for an organization without undermining audience members' capacity to make a rational decision.

In sum, responsible industry environmental campaigns should not use values-based appeals in a vacuum; rather, organizations have an obligation to also provide accurate information about the organization's environmental identity and activities. When using values-based appeals, advertisers must first ensure their adherence to the first two criteria for ethical marketplace advocacy: transparency in the message source and accuracy in describing industry activities. Where time and space requirements are relatively unlimited, such as on websites and in corporate reports, this information can be provided in detail. While a 30- or 60-second ad does impose limitations on the amount of information that can be presented, a values appeal can still be balanced with an accurate statement about the organization's current or proposed initiatives to reduce greenhouse emissions or otherwise contribute to environmental sustainability.

Conclusion

The stakes are high when it comes to industry environmental campaigns. All persuasion requires attention to ethics to ensure target audience members are respected as ends in themselves, and not merely treated as a means to achieving the persuader's goals. For industry environmental campaigns, the environment is also at stake. Public attitudes toward the environment, the environmental crisis, and the need for new policy and regulation can be expected to impact the ability of governing bodies to mitigate the effects of climate change. These stakes call for particular vigilance on the part of communication professionals to ensure industry environmental campaigns are ethical, to avoid charges of greenwashing, and to appropriately communicate genuine industry commitment to the environment.

Despite calls for action in response to climate change, many industry campaigns use tactics such as corporate reports, websites, advertising, and public relations to argue for the status quo in energy policies. Lobbying and public relations efforts in particular are often used to downplay industry's environmental impact while publicizing adverse effects to the economy that may result from environmental restrictions. At the same time, industry media campaigns, especially within the energy sector, frequently tout the positive message that corporate initiatives can provide the solution to environmental problems. And while some businesses are genuinely committed to pro-social and environmental causes, as a group, many corporate environmental campaigns have been described as seeking to "assuage the concerns of the public, deflect blame away from polluting corporations, and promote voluntary measures over bona fide regulation" (Kenny Bruno, quoted by Dahl, 2010).

Industry environmental campaigns are a form of persuasive corporate communication that serves the interests of the sponsoring company. These campaigns can be described as marketplace advocacy, which protects the organization's position in the marketplace by building acceptance for its product and processes, reducing current or potential concerns about risks associated with the industry, and defending it against calls for government regulation. While ethical environmental marketplace advocacy is possible, particular attention is required to avoid the irresponsibility associated with greenwashing and to fulfill the social responsibility to stakeholders that ideally underlies any form of corporate communication.

The criteria of visibility, veracity, and sensitivity are guidelines for creating ethical corporate environmental campaigns. To satisfy the criterion of visibility, industry environmental campaigns must be transparent about the source of the message, particularly when the source is an industry trade association as opposed to an individual corporation. While deceptive "stealth" front groups are patently unethical, even associations without deceptive names have a responsibility to indicate their membership. The criterion of veracity, meanwhile, requires accurate description of the organization's activities and their environmental impact. Deception is unethical, and exaggeration or spin must also be avoided. Organizations must also accurately indicate when their environmental initiatives fulfill the requirements of government mandates or legally binding settlements and not discuss these activities as if they are voluntary.

Finally, the criterion of sensitivity requires ethical corporate environmental campaigns to balance the interests of the organization with social responsibility. Corporate environmental campaigns often focus on scientific innovation and its ability to solve problems. While scientific innovation is rightly valued by society, presenting this value as the sole solution to the environmental crisis is not socially responsible. Such messages may lead the public to set aside their environmental concerns under the assumption that environmental problems will be solved by science and the industry without the need for any other changes. Research findings indicate individuals' environmental concern does not decrease favorability of response to an industry environmental campaign, but is, in fact, positively associated with message acceptance. While values-based messaging in and of itself is not unethical, appropriate advocacy must also honor societal and ethical concerns and avoid undermining audience members' own environmental concern with values-based messages. Ethical corporate environmental campaigns must not overstate the promise of future scientific innovation to solve the current environmental crisis, and values appeals should be balanced with accurate information about the organization's environmental activities.

The example of ExxonMobil's "Advanced Biofuels" ad illustrates some of the pitfalls and true challenges for communication professionals seeking to ethically communicate about industry environmental initiatives. Communication professionals must design effective ads for their clients, and effective advertising should engage emotions and present a clear and relatively simple message. Ethical advertising on this topic, however, requires careful consideration of the intent in engaging emotion, the type of information that is included in the ad, and the possible societal effects of the message. Communication professionals can ultimately best serve the corporations they work for, and the broader society, by considering how industry environmental campaigns will satisfy the ethical criteria of visibility, veracity, and sensitivity.

References

American Coalition for Clean Coal Electricity. (2010). America's power. Retrieved from http://www. americaspower.org/The-Facts/Power-House

ASA. (2011, March 9). Adjudication on ExxonMobil UK Ltd. Retrieved from http://www.asa.org .uk/Rulings/Adjudications/2011/3/ExxonMobil-UK-Ltd/TF_ADJ_49877.aspx

Balmer, J. M. T., Powell, S. R., & Greyser, S. A. (2011). Explicating ethical corporate marketing. Insights from the BP Deepwater Horizon catastrophe: The ethical brand that exploded and then imploded. *Journal of Business Ethics*, 102, 1–14.

Banerjee, S., Gulas, C. S., & Iyer, E. (1995). Shades of green: A multidimensional analysis of environmental advertising. *Journal of Advertising*, 24(2), 21–31.

Bostdorff, D. M., & Vibbert, S. L. (1994). Values advocacy: Enhancing organizational images, deflecting public criticism, and grounding future arguments. *Public Relations Review*, 20(2), 141–158.

Corbett, J. (2006). *Communicating nature: How we create and understand environmental messages*. Washington, DC: Island Press.

Cutler, B. D., & Muehling, D. D. (1989). Advocacy advertising and the boundaries of commercial speech. *Journal of Advertising*, 18(3), 40–50.

Dahl, R. (2010). Greenwashing: Do you know what you're buying? *Environmental Health Perspectives*, 118 (6), A246–A252.

Davis, J. J. (1992). Ethics and environmental marketing. *Journal of Business Ethics, 11*, 81–87.

Edgett, R. (2009). Toward an ethical framework for advocacy in public relations. *Journal of Public Relations Research, 14*(1), 1–26.

Euractive. (2011). Exxon's algae biofuel ad banned over "misleading" claims. Retrieved from http://www.euractiv.com /climate-environment /exxons-algae-biofuel-ad-banned-m-news -503443

Fukukawa, K., Balmer, J. M. T., & Gray, E. R. (2007). Mapping the interface between corporate identity, ethics, and corporate social responsibility. *Journal of Business Ethics, 76*(1), 1–5.

Global Commission on the Economy and Climate. (2014). The new climate economy. Retrieved from http://newclimateeconomy.net

Greenpeace. (2013). Introduction to StopGreenwash.org. Retrieved from http://www .stopgreenwash.org/introduction

Herndon, A. (2013). Exxon refocusing algae biofuels program after $100 million spend. Retrieved from http://www.bloomberg.com /news/articles/2013-05-21/exxon-refocusing -algae-biofuels-program-after-100-million-spend

Homburg, C., Stier, M., & Bornemann, T. (2013). Corporate social responsibility in business-to-business markets: How organizational customers account for supplier corporate social responsibility engagement. *Journal of Marketing, 77*, 54–72.

Hyman, M. (2009). Responsible ads: A workable ideal. *Journal of Business Ethics, 87*, 199–210.

Juhasz, A. (2013). Big Oil's big lies about alternative energy. Retrieved from http://www .rollingstone.com /politics/news/ big-oils-big-lies-about-alternative-energy-20130625

Laufer, W. S. (2003). Social accountability and corporate greenwashing. *Journal of Business Ethics, 43*, 253–261.

Lerner, A. B. (2015). ExxonMobil CEO mocks renewable energy in shareholder speech. Retrieved from http://www.politico.com /story/2015/05/exxonmobil-ceo-downplays-climate-change -mock-renewable-energy-118330

L'Etang, J. (1996). Corporate responsibility and public relations ethics. In J. L'Etang & M. Pieczka (Eds.), *Critical perspectives in public relations* (pp. 82–105). London: International Thomson Business Press.

Martinson, D. L. (1996). "Truthfulness" in communication is both a reasonable and achievable goal for public relations practitioners. *Public Relations Quarterly, 41*, 42–45.

Marty, D. (2007, August 17). Calling all consumers: Green marketing for green products. *E: The Environmental Magazine.*

Miller, B. M., & Lellis, J. (2015). Response to marketplace advocacy messages by sponsor and topic within the energy industry: Should corporations or industry trade groups do the talking? *Journal of Applied Communication Research 43*(1), 66–90.

Miller, B. M., & Sinclair, J. (2009a). Community stakeholder responses to advocacy advertising: Trust, accountability, and the persuasion knowledge model. *Journal of Advertising, 38*(2), 37–52.

Miller, B. M., & Sinclair, J. (2009b). A model of public response to marketplace advocacy. *Journalism and Mass Communication Quarterly, 86*(3), 613–629.

Miller, B. M., & Sinclair, J. (2014). U.S. public response to corporate environmental messages. Paper presented to the annual meeting of the International Communication Association, Seattle, WA.

Plec, E., & Pettenger, M. (2012). Greenwashing consumption: The didactic framing of ExxonMobil's Energy Solutions. *Environmental Communication, 6*(4), 459–476.

Rudinow, J. (1978). Manipiulation. *Ethics*, 88(4), 338–347.

Schlichting, I. (2013). Strategic framing of climate change by industry actors: A meta-analysis. *Environmental Communication, 7*(4), 493–511.

Sethi, S. P. (1977). *Advocacy advertising and large corporations*. Lexington, MA: Lexington Books.

Shellenberger, M., & Nordhaus, T. (2004). The death of environmentalism: Global warming politics in a post-environmental world. Retrieved from http://www.thebreakthrough.org/images/Death_of_Environmentalism.pdf

Sher, S. (2011). A framework for assessing immorally manipulative marketing tactics. *Journal of Business Ethics, 102*, 97–118.

Sinclair, J., & Miller, B. M. (2010). Understanding public response to technology advocacy campaigns: A persuasion knowledge approach. In L. Kahlor & P. A. Stout (Eds.), *Communicating science: New agendas in communication* (pp. 88–108). New York, NY: Routledge.

Sinclair, J., & Miller, B. M. (2012). Public response before and after a crisis: Appeals to values and outcomes for environmental attitudes. In Lee Ahern & Denise Bortree (Eds.), *Talking green: Exploring contemporary issues in environmental communications* (pp. 107–130). New York, NY: Peter Lang.

Terrachoice. (2009). The seven sins of greenwashing: Environmental claims in consumer markets. Summary Report, North America. Retrieved from http://sinsofgreenwashing.com/findings/greenwashing-report-2009/index.html

Union of Concerned Scientists. (2014). Exposing the disinformation playbook: An interactive slide show. Retrieved from http://www.ucsusa.org/global_warming/solutions/fight-misinformation/global-warming-facts-and-fossil-fuel-industry-disinformation-tactics.html

West Virginia Coal Association. (2009). Coal facts 2009. Retrieved from http://www.wvcoal.com/docs/Coal%20Facts%202009.pdf.

9

Scientists' Duty to Communicate: Exploring Ethics, Public Communication, and Scientific Practice

SARAH R. DAVIES

Why should scientists communicate to public audiences? There are lots of possible answers to this. It can be (perhaps should be) enjoyable, for the scientists themselves and for those they are interacting with (journalists, laypeople, schoolchildren). Scientists (as well as their audiences) might learn something new, or get help in putting their work into a wider context. They can inform people about aspects of science that are unfamiliar to them or inspire them to take a greater interest in science.

All of these motivations for science communication have been charted in the literature (see, e.g., Horst, 2013). But intertwined with these rather personal drives often comes something that is more normative. There is a sense, both from scientists themselves and from more general discussions of science communication, that researchers have a duty to communicate their work. Science communication is, in this understanding, one of the responsibilities that scientists take on when they are funded by the taxpayer (as they often are) to carry out scientific research.

It is this sense of science communication as a duty, or responsibility, that I want to reflect on (and perhaps problematize) in this chapter. In a book that looks at the ethics of science communication, I will consider the ethical responsibilities that lie at the heart of scientific practice itself. Is science communication integral to what we expect of our publicly funded scientists, part of the implicit social contract between science and society? What reasons are there for thinking of it in these terms—and why might we disagree with such a position?

This is not going to be an abstracted or especially philosophical discussion. I want to focus on the views and experiences of scientists themselves, and use these to explore the responsibilities of science to society. Though I will start by discussing the different arguments that have emerged over the last decades about how and why scientists should communicate, I will quickly move on to the voices of scientists themselves. In the next section, I will briefly consider previous research that has asked scientists about their views on public communication, while in the following one I use data from my own research to dig down further into what scientists view their responsibilities as being. The final section attempts to draw all this together and to reflect on whether we should see public communication as a central ethical duty of scientists. My answer is yes—but that meeting this duty is more complex than the framing of the question might imply. There are good reasons for viewing science communication as part of science's responsibility to society, but we also need to consider the challenges of contemporary scientific working practices and conditions. As we'll see, there is more at stake for scientists than a simple choice between communicating or not communicating.

Public Communication as Scientific Responsibility: The Story so Far

Scientists have thought about how to communicate with the public since science first began (Shapin, 1990). However, a convenient place for this account to start is some 30 years ago, with the publication of the UK Royal Society's Report on the Public Understanding of Science[1] (Royal Society, 1985). This is often described as a seminal moment in science communication because it placed scientists' public communication activities firmly on the political agenda (initially in the United Kingdom, but then also around the world; Gregory & Lock, 2008). It ends with a call to action. After a long series of recommendations, the final sentence runs as follows:

> Our most direct and urgent message must be to the scientists themselves: Learn to communicate with the public, be willing to do so and consider it your duty to do so. (Royal Society, 1985, p. 36)

According to the Royal Society report, public communication is a central *duty* that all scientists need to accept. Earlier on the authors explained why this is the case. Public understanding of science is important for all kinds of reasons ranging from the cultural (laypeople should be able to appreciate science and its impacts on society) to the economic (a technological society requires people who can use and develop its technologies).[2] Thus far—the

---

1. Also known as the Bodmer Report, after its lead author.
2. See Durant, Evans, & Thomas (1989) for a similar account of why scientific literacy is important for the general public.

report argues—scientists have tended to rely on other people to ensure that the public understands their work—people like a few key, visible scientists (see Goodell, 1975) or science writers and journalists. But this is no longer seen as appropriate. For one, it's better to go straight to the horse's mouth: scientists are always going to understand their own research best (and should be prepared to write lay summaries of it, for instance). But there's also a question of accountability. Scientists are "democratically accountable to those who support scientific training and research through public taxation" (Royal Society, 1985, p. 24). Anyone who pays taxes has a right to know what their money is going toward and to be assured that it's being used wisely; therefore, any publicly funded scientist should be ready to explain their work and why it's valuable. Somewhat less idealistically, the report also points out that if laypeople don't know what scientists are up to, they're unlikely to care whether politicians cut science funding.

The notion of democratic accountability has continued to be a key rationale for scientists' involvement in public communication over the past decades. Discussion of this involvement has, however, shifted focus somewhat, with an increasing emphasis not just on communication—scientists being able to explain their science to public audiences—but also on dialogue. A 2009 US report on public *engagement* with science (note the change in language) argued that engagement activities are "about allowing a greater segment of society to take ownership of and help direct scientific investment and application" (McCallie et al., 2009, p. 28). Democratic accountability in this context is not just enabling the public to hear about the research being carried out with their money, but allowing them to "take ownership" of that research. Similarly, a communication guide for scientists from the United Kingdom (titled *The Engaging Researcher*) says that engagement is "a two-way process, involving interaction and listening, with the goal of generating mutual benefit," and that accountability is a key reason behind it (Duncan & Spicer, 2010, p. 4). In contemporary discussions of science communication the duty of scientists to account for their research is not met solely by participating in activities like writing a newspaper article, giving a lecture, or speaking to a journalist. Scientists are expected to engage as well as to disseminate—or, in other words, to listen as well as to speak. Science as a whole should be *responsive* to public views and concerns.

Importantly, these trends are being actively promoted by funders of research—those who siphon money from taxpayers along to scientists. If you are a scientist or researcher (social scientists and humanities scholars are asked to communicate, too) and have applied for research funding in recent years, you will doubtless have been asked to explain how you will disseminate your work, ensure its impact on society, or engage with public stakeholders. In Europe, one of the key expectations of research funded by the European Commission (for instance, in its Horizon 2020 program[3]) is a commitment to responsible research and innovation, or RRI. RRI is something that researchers are asked to sign up to (metaphorically if not literally), and involves a set of six "keys" that includes ethics—understood as carrying out research that respects "fundamental rights"—and engagement, or ensuring that "all societal actors" can participate in research and innovation (European Commission, 2013). While RRI is being presented as a framework for research in Europe and beyond (many US scholars are also promoting the notion[4]), at the moment it is not entirely clear what it should mean for individual scientists. As RRI scholar Bert-Jaap Koops has recently pointed out, despite the term being increasingly used it is rather hard to pin down exactly what it is or how it should be carried out (Koops, 2015). But it is certainly indicative of a wider trend of expecting scientists to listen to their publics, and, in listening, to take responsibility for responding to their concerns. The one thing all discussions of RRI have in common, Koops notes, is an emphasis on public engagement with research. "In line with the 'participatory turn' in the social sciences," he writes, "responsible innovation researchers emphasize the importance of listening to stakeholders in innovation processes" (Koops, 2015, pp. 5–6).

Scientists on Public Communication: Motivations and Drivers

Many discussions of how scientific research should be funded and carried out have thus emphasized the need for scientists to be open, publicly accountable, and responsive with regard to their work. Public communication—whether one-way or dialogic in form—has increasingly been seen as integral to the practice of publicly funded research, and a duty of the beneficiaries of such funds. By 2001, science communication scholar Steve Miller could write that two key effects of the 1985 Royal Society Report had been to legitimize science communication as an acceptable activity for working scientists, and to mobilize the scientific community to carry out that communication (Miller, 2001). Support for enhancing public understanding of science had meant, he wrote, both that money was increasingly available to help scientists

---

3. Horizon 2020 is the framework program for European research funding between 2014 and 2020. See http://ec.europa.eu/programmes/horizon2020/en/what-horizon-2020

4. See, for instance, the US-based *Journal of Responsible Innovation*: http://www.tandfonline.com/loi/tjri

carry out communication activities and that scholars had access to training in such activities. It seems that science communication has become normalized: at my own institution, the University of Copenhagen, for instance, the "Responsible Conduct of Research" course that all PhD students must take includes training in public communication alongside discussion of research misconduct, fraud, and conflicts of interest.

But what are scientists themselves making of all this? In exploring the practice of contemporary science communication, researchers have charted some of the experiences that scientists are having—including the fact that, in contrast to Miller's optimism, while scientists may be familiar with a "duty" to communicate, they often continue to see it as a nonessential and even distracting activity (Casini & Neresini, 2013; Horst, 2013; Royal Society, 2006). More recent terms such as RRI fare even worse. Many scientists are unaware of these policy discussions, and (once informed) struggle to see their relevance to their everyday activities (Davies & Horst, 2015a; Kjølberg & Strand, 2011). Even where scientists are enthusiastic about science communication, there is generally a strong sense that there are many practical barriers to carrying it out, from time pressures to disapproval from colleagues or lack of funding (Davies, 2013a; Royal Society, 2006).

More positively, it seems that scientists and audiences alike tend to get a great deal of pleasure from participating in science communication activities, and that scientists, having once participated, are often keen to continue to be involved (Besley, Oh, & Nisbet, 2013; Davies, 2013a; Pearson, Pringle, & Thomas, 1997; Poliakoff & Webb, 2007). Motivations for involvement include increasing public understanding and appreciation of science, inspiring young people to pursue scientific careers, raising the profile of a research center or institute, being encouraged (or forced) by a colleague, career development, and personal satisfaction and enjoyment (Davies, 2013b; Martín-Sempere, Garzon-Garcia, & Rey-Rocha, 2008). Particularly pertinent to the discussion here is that a sense of duty also seems to play a role in motivating scientists to carry out science communication. Many accept that it is an "obligation": in one UK study, when researchers were not carrying out public communication they sometimes expressed some awkwardness or guilt about this, as well as the sense that it was something one "should be doing" (Davies, 2013a). How this duty is understood and expressed may vary between researchers. Martín-Sempere and her coauthors (2008) suggest that "duty" as a motivation for engaging in science communication is primarily expressed by more senior researchers (as opposed to, say, postdocs and technical staff). And Horst, in a discussion of what scientists see themselves as representing when they communicate science (their institution, their discipline, science as a whole) notes that there may be a difference between a "sense of duty to make expert knowledge available," on the one hand, and a duty "to increase enlightenment and rationality in society," on the other (2013, p. 773). Though scientists often acknowledge that they have a responsibility to communicate their work to society, then, how this is expressed and worked out in practice (if at all) seems to vary.

Being Responsible in Scientific Practice

In this section I want to reflect on findings from recent research (carried out by myself and my colleagues) in order to further consider how scientists think about and experience science communication as a duty they need to fulfill. The data I'm drawing on comes from interviews with PIs (principal investigators[5]) and research group leaders in Denmark, the United Kingdom, and the United States; the aim was to focus on the "meso-level" of research management in order to complement other parts of the project that looked at practices at the bench and at the level of policy making (see Davies & Horst, 2015b; Glerup, 2014; Glerup & Horst, 2014). As a whole, the project explored how discussions of responsibility in science—such as that around RRI— are being interpreted and put into practice by those working in science. In these interviews (29 in total, spread across the three countries and a range of natural science disciplines; further details are found in Davies & Horst, 2015a) we asked about interviewees' biographies, their roles and responsibilities, and their awareness of different kinds of demands for "responsible research" from policy makers and funders. The conversations we had with these research managers and leaders therefore sought to understand their perspectives on the responsibilities and duties of working as a scientist. What did they see their responsibilities as being—and how did this relate to policy demands for RRI or other kinds of responsive science (Owen, Macnaghten, & Stilgoe, 2012)?[6]

The first point I want to make is that though the scientists we spoke to

5. Principal investigator, or PI, generally refers to the lead researcher on a monetary grant or contract that funds scientific research. However, the term is sometimes used to refer to anyone who leads a research team.

6. In my discussion here I am not giving a full analysis of this dataset, but instead giving the reader a sense of the key themes that emerged from it and the ways in which these relate to science communication as an ethical obligation within research. Fuller analyses can be found in Davies & Horst (2015a). In what follows I pay particular attention to instances where interviewees reflected on the ethics of the science-society relationship; throughout, the quotes given are illustrative of broader themes in the data. All names have been changed.

tended to have little familiarity with recent policy calls for "responsible re-search" and dialogue (cf. Kjølberg & Strand, 2011), this didn't mean that they saw themselves as irresponsible or as having no ethical obligations. Quite the reverse; almost universally, our interviewees had rather finely honed views on what "good" science—meaning ethical science—should look like. And rather than depicting themselves as footloose and fancy-free, able to delegate moral concerns to other actors (a division of labor that has previously been charted; Shelley-Egan, 2010), they instead talked about being, at times, al-most overburdened with responsibilities. Scientific work, it was clear, is ex-perienced not as an ethically neutral practice but as something involving a complex web of rights, duties, and responsibilities.

Not surprisingly, many of these responsibilities and duties related to the science itself. The PIs, group leaders, and managers we spoke to saw them-selves as being responsible for the quality of the science that their lab pro-duced. As such, they had to work both to monitor its progress (keeping in touch with experiments, analyzing data, reading draft papers) and to place it into its wider scientific context (keeping up to date with the field, keep-ing track of collaborators and competitors, identifying new research direc-tions). They were ultimately responsible for both the day-to-day concerns that kept their research group running—and thereby producing robust sci-entific results—and the broader imagination of where their science could and should be going. Many that we spoke to were not content to simply keep things ticking along. Instead, as in the quote below, they were concerned to "make a contribution":

> My role is to lead, advise, direct, keep the overall direction of our work mov-ing forward . . . making sure that we're productive and responsive to our fund-ing agencies, helping to ensure that we are really generating new knowledge. I'm not satisfied with nibbling around the fringes of the issues. I want to make sure that what we're doing is really making a contribution.

The speaker, Tara, is a full professor at a US university with a group of some 13 people. She saw her role, she told us, as being "like running a little company." She was responsible for strategy and direction and for bringing in resources to keep the science happening. But this was not just in order to keep her group in gainful employment. She was clear that it was important to be "really generating new knowledge," doing something scientifically mean-ingful, and contributing to the intellectual development of her field. Many of her duties, then, related to ensuring that her group's science was funded, rigorously carried out, and thoughtfully directed. In this sense she had an obligation to the science itself. It was her duty—and indeed, her personal desire—that it was the best that it could be, not something that was just "nibbling around the fringes of the issues."

Another set of responsibilities related to the people who carried out that science. Even the more junior scholars we spoke to (those working at an assistant professor level or the local equivalent) saw themselves as having a duty of care to the students, technicians, postdocs, and other scientific staff who worked with them.[7] They had to find them good projects, make sure they had a productive and supportive working environment, teach them the skills needed for the science at hand and for scientific life in general, guide them through the frustrations of failures or negative results, and help them secure jobs or career stability. Several of those we spoke to depicted themselves as the "parents" of their groups. It was their role, they told us, to be the "breadwinner," help their colleagues through personal and professional difficulties, and smooth interpersonal conflicts. Sometimes that might mean being harsh—they might have to rebuke a student for relying too much on the help of a technician, or explain that someone's work veered dangerously close to plagiarism. But all of this had to be done with the good of group members in mind, not the PI's own success or reputation. It was the PI's responsibility to look after their group, and to try to ensure, as one scientist told us, that individuals are "happy and flourish."

The extent to which our interviewees felt responsible for the well-being of their group members was striking. Leif, for instance, was a senior Danish professor. When we asked him about the responsibilities his role involved, he was quite clear that they were first and foremost to people:

> That's to people, that they have fun, being here. That's—I guess that's my responsibility. And it's not my responsibility, but it's part of my responsibility, that when young PhD students, they are finishing here, they have something to do. So we don't make them so specialized that they have no chance to go anywhere . . . if I see somebody that is teaching too much, and they are really close to having stress or something like that, then I should know before they know.

Leif mentions several ways in which he is responsible for his team: he wants to make sure that they "have fun" working in his institute (he leads a large group that includes several others at the full or associate professor level); it is partly his responsibility (shared with the students themselves) that students "have something to do" once they complete their PhD, through ensuring that their training is not so specialized that they are unemployable

---

7. See Davies & Horst (2015a) for a longer discussion of the role of care in research management.

elsewhere; and he tends to their mental and physical well-being by monitoring their workload and trying to "know before they know" whether someone is becoming stressed. Many others said something similar. Though expectations around the appropriate level of personal and pastoral support varied, almost all the scientists we spoke to saw such support as one of their primary responsibilities. As with Leif, this was often the first thing they spoke about as they reflected on their roles and duties: they talked about caring for their group before turning to other obligations they felt they had (such as to their science, funders, or institution). Indeed, being responsible for individuals was often closely related to the responsibility to ensure that the group was producing good quality science. Most of our interviewees thought that good science was not produced by unhappy groups. Ensuring that group members were comfortable, confident, and happily collaborating therefore meant that the PI was not just fulfilling their responsibilities to their colleagues but also to the science itself.

We seem to have come rather a long way from public science communication as a duty of scientific life. This is because, at least at first, the PIs and group leaders we spoke to simply did not talk about this as one of their responsibilities. As noted, they tended to turn straight to talking about the duty of care they had to their group and, through that, to the science they were producing. This was their priority, and, in most cases, the primary space in which their everyday work took place—the area, in other words, where they were most intimately involved and where their actions could have the most tangible effects. But as we asked them more about their roles and responsibilities, other kinds of duties and ethical commitments started to emerge, from wanting their work to reflect well on their institution to writing fair reviews on articles or being committed to open access publishing. As I have said, it was clear that science was seen as a moral endeavor and that there were good and bad ways of doing it.

Many of these moral commitments related to communication. Steve, for instance, was a relatively junior UK physicist who was quite clearly interested in interacting with the world outside of his immediate colleagues. He sat on committees in his department, was involved with the UK's Institute of Physics, and worked with the UK's Athena SWAN and Juno Projects (which respectively seek to increase the representation of women in science generally and physics specifically). A responsible scientist, he said, should be engaged with the "running of science." Outside of his responsibilities to his students and colleagues, he was clear that there was a wider "research ethics" that he was committed to and which involved not only the "obvious things" like not falsifying data but behaviors like publishing truthfully (not, for instance, giving the impression that a particular experiment was easy to carry out when it wasn't or avoiding writing about negative results) and responding to questions, comments, and criticism on published work (what he called "providing an after-sales service"). Although Steve—like many others—had some criticisms of the way in which UK funders were increasingly asking for assessments of his work's "impact,"[8] he was also sympathetic to what he saw as the logic behind these demands: "If you're getting government grants you're spending public money, and you should do it wisely":

> I ought to justify to the public why I am doing my science. And so when I go to a party and I meet people, they say what do you do? I'm a physicist. What are you researching, da da da, and what are the applications? That's what always comes out, is what are the applications. Well, I should answer what is the impact of my research. . . . I should be able to tell someone I meet at a party why 0.1 pence of their tax that year went on my project. . . . The other responsibility is to society to do stuff that is not just playing.

This, of course, is the argument that we began the chapter with. If scientists are publicly funded, then they have a responsibility to be able to account for the ways in which they are using those public funds. Steve was interested in outreach and public engagement—he had developed a science round for a pub quiz his parents were running, and wanted to get more involved in outreach activities—but here he uses the example of going to a party and being asked what he does. He should, he says, be able to talk about his work, including its "applications" and "impact"; ultimately, he needs to be able to explain to his interlocutor why "0.1 pence of their tax . . . went on my project." Beyond this, however, he also needs to be doing research that "is not just playing." If he is to be able to genuinely account to others for the public value of his work, then he himself needs to be convinced of its public value—that it is, in other words, scientifically (and, perhaps, eventually industrially) useful.

Others offered similar reflections on how their work did and should relate to wider society. Was their research "such that the public will not be unhappy about it"? How could they help equip the general public for a world that will increasingly be "dictated by science"? At what point should they decide that their research "is solid enough to be presented" to wider audiences? How can members of the public best access scientific results, given that "they pay for it . . . it's their knowledge"? These kinds of questions were often tied to implicit, intuitively developed ethical systems that viewed science as responsible to, and dependent on, wider society. Though interviewees might argue

8. See Martin (2011) for a discussion of the issues involved.

passionately for the value of basic research, or note that they themselves found public communication difficult, they almost universally articulated a sense of duty toward the societies that funded them. Like Steve, they didn't feel it was appropriate to be "just playing." Their work wasn't a hobby, but a profession that should be done to the best of their ability and which should have intellectual and societal worth.

Truthfulness, or honesty, was an important part of this implicit ethics. This certainly applied to public communication, but it was rather difficult, in the talk of our interviewees, to disentangle this form of scientific communication from others. "Over-hyping," for instance, was bad whether it was to funders, the media, or one's disciplinary community. As we have seen, Steve's personal "research ethics" included a sense of how to publish: One should write honestly, mentioning the flaws of one's work as well as its successes. Similarly, Hanna, a PI based in Denmark, spoke about the "rules" she followed as she applied for funding for her research:

> I mean, I've heard people say things like they write proposals where they've already done half the work and I don't do that. I sort of do what I say I'm going to do. I'm not, you know—I'm sort of a bit of a stickler for rules and these sorts of things. So I don't play those sorts of games.

Hanna, then, was truthful in her interactions with funders, rejecting the "games" that she felt were common in the constant hunt to gain research funds. More generally, she took transparency as a key principle. Like Steve, she endeavored to publish in a way that was honest, not "over-hyping things to try and push them into a better journal," and she worked hard to ensure that all contributions to the work, whether from her group or beyond, were recognized ("authorship," she said, had to be "transparent"). Indeed, this sense of honesty was one reason that she wasn't as involved in public outreach and debate as she might have been. Working in the potentially controversial area of nanoscience, and aware that there was a "public debate about safety issues," she nonetheless did not get much involved in this because, she said, "I don't feel like I'm qualified to comment on it." Instead she focused on educating the students that she worked with—in the science, certainly, but also in a way of doing science that she felt was ethically robust.

Negotiating "Tiers of Responsibilities"

What we've seen so far, then, is that scientists—or at least those working at the group leader level—do tend to have a set of ethical commitments that they mobilize in making sense of their working lives. They feel responsible for their colleagues, in particular for those working in the group that they lead, and for the science that they produce. More broadly, they also have a sense that they are accountable to the societies that fund them. Just as previous research has indicated (Horst, 2013; Martín-Sempere et al., 2008), and as the Royal Society Report argued back in 1985, scientists feel a duty to be able to explain, and perhaps even discuss, their research with the taxpayers who fund it. As well as there being good theoretical arguments for researchers to publicly communicate their work—that, in a democracy, citizens should know about the activities of the state (Durant et al., 1989)—scientists themselves feel a duty to be publicly accountable. For those we spoke to, science is not an autonomous entity, separate from the rest of society and subject to no rules but its own, but is intertwined with the state and its citizens. Public funding implies a responsibility, and an openness, to public audiences.

However, in closing this chapter I want to complicate this narrative somewhat. The story is not quite so straightforward: It is not always clear how scientists can or should act on the duty they have to communicate. Yes, scientists do have a responsibility toward their paymasters (ultimately, the public)—but, from what interviewees told us, what this means in practice will look different from case to case.

Hanna's views, discussed above, are a useful starting point in showing how personal ethical systems might in practice be more complex than sets of "duties" to be fulfilled in all circumstances. Hanna was clearly aware of public debate related to her work in nanoscience. There were discussions, she said, she could have contributed to, but she didn't "engage in much of a public debate on these types of things," because even though "on paper" it might look as though she had relevant expertise, "I don't feel like I'm qualified to comment on it." Her responsibility to treat the science well, to respect her limits and not run the risk of pontificating on something she did not fully understand, was greater than her responsibility for public communication and debate. Internally, and possibly not even consciously, Hanna had performed an assessment and decided that her priority should be education and her own ethical practices in science. It was more useful for her to invest her time and energy in creating an environment for her students where transparency, honesty, and generosity were the norm, and where, as she said, there was no pressure to "cut corners" in their research, than it was for her to try to influence public debate or policy. For her, at least at this particular moment, the ethically correct behavior was to *avoid* public communication. Another group leader noted that though they did some outreach, they did not feel that they were very good at it; similarly, then, they felt that it should not be a personal priority.

What this starts to point to is something that everyone we spoke to told us: scientists (and particularly those at the managerial level) are busy. Our interviewees managed their labs and their students, taught undergraduates, kept in touch with industrial and academic collaborators, went to meetings and conferences, organized department research seminars, engaged with policy makers and funders, sat on university committees, examined PhD theses—and, of course, participated in the seemingly never-ending search for research funding. Inevitably, they had to manage and prioritize the various duties and responsibilities that they felt they had. Their experience was of "different tiers of responsibilities" as Tara, the US professor quoted earlier, put it. These tiers might cover their group, their undergraduate students, their institution, local taxpayers, their field as a whole, and society at large (for instance).

As I noted earlier, the first things that came to mind as interviewees thought about these tiered responsibilities were their science and their group (the two being interlinked). These were the things they felt most responsible for, and where their attention from day to day was focused. Public communication, and being publicly accountable, even though understood as a duty they should take on, was often something that lay rather far from the immediacy of most of their work, and at the outer edge of their activities and resources. Even though scientists might say, as one PI did to us of outreach, "we probably could do and should do more," this was generally easier said than done. It would require reaching outside of the familiar (if always challenging) structures of the group and the discipline, into spaces where agency is not so easily exerted and where it is not so clear what obeying a duty of care would look like. Simply put, most scientists felt that they had to prioritize acting within their immediate sphere of influence. Their responsibility to the public was not so urgent, nor so tangible, as their responsibilities to their students.

Ulster, a professor in the United Kingdom, explained this particularly clearly. As he listed the different "tiers" of responsibilities he had (some details of which I have cut in order to keep the quote concise) he was clear that his priority had to be his immediate team:

> More than anything else I feel like as a supervisor my responsibility is to my team to make sure that they can do their work, they succeed in their work and they can go on and find jobs of their own elsewhere. I feel responsibility to the grants that I've got, to make sure that work gets done on those. . . . I feel a bit of responsibility to the subject itself and [the university] to try and deliver things that will have a benefit in general, economically or for the good of the world in some way, rather than just necessarily doing research for research's sake . . . but the weight of responsibility that I feel day to day for that is significantly less than it is for making sure that the team, the individuals around me,

do well and don't fall into a hole where their results never work and no one's looking after them.

Ulster, then, does have a sense of responsibility to "the world." He's concerned both to account for his research activities to his funders and to try to work on scientific questions that will have some "benefit," economic or otherwise. Like Steve, he doesn't think it's appropriate to do "research for research's sake." But he is also clear that these responsibilities are more diffuse, less relevant to his everyday experience, than those to his team. "More than anything else," he says, his responsibility is to make sure that those in his group succeed in their work, and that they're being "looked after." The "weight of responsibility" that he feels from day to day is always based on that.

In seeking to negotiate the different claims and responsibilities that lay on them, Ulster and others that we spoke to were engaged in a constant balancing act. It was not that they did not care about public communication, or about dialogue or policy terms like RRI; when we spoke about these things in the interviews, scientists were often interested and supportive. (After all, it is hard to argue against responsible science. Who wants the other kind?) But they also felt that other duties and concerns were more immediate and pressing. It was hard enough for them to care adequately for their students and staff, and to give them the time they deserved. To find time for public engagement, or outreach, or dialogue, took a special kind of commitment and personal drive. Some people had this, certainly, but others, like Hanna, had decided that for them the right thing to do was to invest their energies elsewhere.

Conclusion

This tension—between interest in science communication and other demands on one's time—is a challenge that those calling for and funding public communication and dialogue, as well as scientists themselves, need to reflect on. Most scientists would seem to agree that science has an obligation to society, and that they work on behalf of the societies that fund them, rather than for their own entertainment (as one of our interviewees said, "They don't give you money so that you can satisfy your own curiosity and impress your colleagues"). Most would similarly acknowledge that they have a duty to communicate their work to public audiences, whether that is at a party, a university open day, or a dialogue event. But they are also clear that in practice they may have higher priorities, such as tending to the well-being of the group. The interest in public communication expressed by the scientists we spoke to was therefore not quite along the lines envisaged by frameworks such as RRI or public engagement. As we saw at the start of the chapter, such frameworks have in recent years tended to emphasize science communication as something that is active, dialogic, and responsive—"interaction and listening," as Duncan and Spicer write, "with the goal of generating mutual benefit" (2010, p. 4). The view we were given of public communication by scientists was of a rather more passive responsibility, one that could be readily crowded out by other demands on one's time and ethical energies.

This suggests that calls for increased public engagement and communication by scientists need to be rooted in an understanding of the working practices and conditions of scientists. As I have written, all of those we spoke to described themselves as busy, and at times as oppressively so. Many mentioned the need to "publish or perish" in the context of their students or themselves; some explicitly said they wished they—or their team—had more time to devote to public communication, dialogue, and outreach. Calls for more public communication or dialogue therefore perpetually run the risk of placing further demands on researchers without providing appropriate resources, whether of time, money, or training. A number of commentators have discussed the contemporary pressures of "academic capitalism" on researchers of all kinds (Hackett, 2014). If we are to talk of ethics, then, we should devote some attention to the conditions under which contemporary science is carried out. What is demanded of publicly funded researchers, and how do these demands interact with researchers' own ethical systems? What systemic barriers are there to scientists acting on their interests in, or commitments to, public communication, engagement, and dialogue? How do the conditions of contemporary science limit or enable particular (ethical) behaviors and practices?

What we have seen in this chapter is that researchers will inevitably be balancing a web of responsibilities and duties—to their group, the science, the discipline, society—and that how this plays out will vary from individual to individual. Supporting public engagement may therefore be a question not only of reminding scientists that they have a duty to communicate their research—à la the 1985 Royal Society report—but also of ensuring that their working conditions enable them to do so. For those we spoke to, the will to communicate certainly existed. A greater challenge may be to change science's systems of reward, promotion, and time allocation to enable scientists to more often act on their duty to communicate and deliberate with their public audiences.

References

Besley, J. C., Oh, S. H., & Nisbet, M. (2013). Predicting scientists' participation in public life. *Public Understanding of Science, 22*(8), 971–987.

Casini, S., & Neresini, F. (2013). Behind closed doors: scientists' and science communicators' discourses on science in society: A study across European research institutions. *Technoscienza: Italian Journal of Science & Technology Studies, 3*(2), 37–62.

Davies, S. R. (2013a). Constituting public engagement meanings and genealogies of PEST in two U.K. studies. *Science Communication, 35*(6), 687–707.

Davies, S. R. (2013b). Research staff and public engagement: A UK study. *Higher Education, 66*(6), 725–739.

Davies, S. R., & Horst, M. (2015a). Crafting the group: Care in research management. *Social Studies of Science, 45*(3), 371–393.

Davies, S. R., & Horst, M. (2015b). Responsible innovation in the US, UK and Denmark: Governance landscapes. In B.-J. Koops, I. Oosterlaken, H. Romijn, T. Swierstra, & J. Van den Hoven (Eds.), *Responsible innovation 2* (pp. 37–56). Springer International.

Duncan, S., & Spicer, S. (2010). *The engaging researcher: Inspiring people to engage with your research.* Cambridge, UK: Vitae. Retrieved from http://www.vitae.ac.uk/researchers/260911/Public-engagement.html

Durant, J., Evans, G., & Thomas, G. P. (1989). The public understanding of science. *Nature, 340*, 11–14.

European Commission. (2013). Options for strengthening responsible research and innovation: Report of the expert group on the state of art in Europe on responsible research and innovation. Retrieved from https://ec.europa.eu/research/science-society/document_library/pdf_06/options-for-strengthening_en.pdf

Glerup, C. (2014). Organizing science in society: The conduct and justification of responsible research (Doctoral dissertation, Frederiksberg: Copenhagen Business School).

Glerup, C., & Horst, M. (2014). Mapping "social responsibility" in science. *Journal of Responsible Innovation, 1*(1), 31–50.

Goodell, R. (1975). *The visible scientists.* Boston, MA: Little, Brown.

Gregory, J., & Lock, S. J. (2008). The evolution of "Public Understanding of Science": Public engagement as a tool of science policy in the UK. *Sociology Compass, 2*(4), 1252–1265.

Hackett, E. J. (2014) Academic capitalism. *Science, Technology & Human Values, 39*(5), 635–638.

Horst, M. (2013). A field of expertise, the organization, or science itself? Scientists' perception of representing research in public communication. *Science Communication, 35*(6), 758–779.

Kjølberg, K. L., & Strand, R. (2011). Conversations about responsible nanoresearch. *NanoEthics, 5*(1), 99–113.

Koops, B.-J. (2015). The concepts, approaches, and applications of responsible innovation. In B.-J. Koops, I. Oosterlaken, H. Romijn, T. Swierstra, & J. Van den Hoven (Eds.), *Responsible innovation 2* (pp. 1–18). Springer International.

Martin, B. R. (2011). The research excellence framework and the "impact agenda": Are we creating a Frankenstein monster? *Research Evaluation, 20*(3), 247–254.

Martín-Sempere, M. J., Garzon-Garcia, B., & Rey-Rocha, J. (2008). Scientists' motivation to communicate science and technology to the public: Surveying participants at the Madrid Science Fair. *Public Understanding of Science, 17*(3), 349–367.

McCallie, E., Bell, L., Lohwater, T., Falk, J. H., Lehr, J. L., Lewenstein, B. V., . . . Wiehe, B. 2009. Many experts, many audiences: Public engagement with science and informal science education. A CAISE Inquiry Group Report. Washington, DC: Center for Advancement of Informal Science Education (CAISE). http://caise.insci.org/uploads/docs/public _engagement_with_science.pdf

Miller, S. (2001). Public understanding of science at the crossroads. *Public Understanding of Science*, *10*(1), 115–120.

Owen, R., Macnaghten, P., & Stilgoe, J. (2012). Responsible research and innovation: From science in society to science for society, with society. *Science and Public Policy*, *39*(6), 751–760.

Pearson, G., Pringle, S. M., & Thomas, J. N. (1997). Scientists and the public understanding of science. *Public Understanding of Science*, *6*(3), 279–289.

Poliakoff, E., & Webb, T. L. (2007). What factors predict scientists' intentions to participate in public engagement of science activities? *Science Communication*, *29*(2), 242–263.

Royal Society. (1985). *The public understanding of science*. London: The Royal Society.

Royal Society. (2006). *Survey of factors affecting science communication: Conclusions, recommendations and actions*. London: Royal Society.

Shapin, S. (1990). Science and the public. In R. C. Olby, B. N. Cantor, J. R. R. Christie, & M. J. S. Hodge (Eds.), *Companion to the History of Modern Science* (pp. 990–1007). London: Routledge.

Shelley-Egan, C. (2010). The ambivalence of promising technology. *NanoEthics*, *4*(2), 183–189.

PART THREE

Case Studies

What is a case study? While many of the other chapters in this collection use concrete examples to illustrate more general points, the five in this part go into more depth about specific situations and events that allow them to high-light a range of ethical issues. The focus of a case study is generally on the case itself, wherever it takes us, rather than on a strongly predetermined research question. This is not to say that the choice of a case to study is arbitrary—on the contrary, these authors have all chosen cases that they knew would allow them to explore interesting territory and expand our theoretical understanding through examining concrete social situations. But the central topic that anchors each of these discussions is a specific real-world sequence of events.

Case study researchers can perhaps be likened to historians who generally use whatever information is available to reconstruct and explore events or situations, whether the data are qualitative or quantitative or archival. Our case studies here, of course, focus on the science communication aspects of the events they are considering and analyze the questions these raise with respect to ethics.

This part begins with work by Daniel McKaughan and Kevin Elliott (chapter 10), who note the tension underlying our expectations for scientists when they communicate: We expect them to offer their guidance to other members of society, but we also expect them to be objective about the scientific facts. These two expectations can be in conflict. As a case in point, they consider research on the social-sexual behavior of voles. As it turns out, voles (small rodents who look like hamsters) produce two neuropeptides that affect their pair-bonding behavior—and these may also influence behavior in humans. But does this mean that humans are really all that much like voles? The authors propose that journalists could present research-based information in a way that better reveals to their readers which elements represent interpretations and which should be understood as objective science.

Next up, Lora Arduser (chapter 11) considers another case involving our biological heritage: the growing commercial enterprise of genetic testing marketed not to healthcare professionals, but directly to consumers. Will this empower people to better manage their own health or only mislead them? Arduser examines key documents in the case from a genetic testing company and the Food and Drug Administration plus over 70 consumer comments about the company's service that appeared on amazon.com. Themes of understanding, empowerment, safety and innovation, reliability and accuracy, and privacy dominated this discourse. But some of the things that seemed to generate the strongest consumer concerns may actually surprise you.

Kelly Bronson (chapter 12) then addresses the other side of the biotech coin: agricultural biotechnology. "Ag biotech" has been contested for a variety of reasons for a long time, with public concerns initially most visible in Europe but seemingly on the rise in the United States now as well. Canada, which has tended to be somewhere in the middle, is the site of Bronson's case: farmers in Saskatchewan whose resistance was dismissed by some as stemming from simple ignorance. Her message here is that not all resistance should be understood as "anti-scientific" or "irrational," phrases that can be used to discredit, dismiss, and exclude oppositional voices. Rather, anti-biotech farmers' concerns are best understood as a reflection of their social and environmental values coming in conflict with what to them appeared as "rampant corporate power."

Understanding ethics in practice requires thoughtful consideration of "meanings-in-use," argues Alain Létourneau (chapter 13), who also uses a Canadian plant science example—but a very different one. What people actually mean by "ethics" can vary widely, be specific to the context, and diverge from the definitions of scholars. Various ethics-related values and norms, in practice, can conflict. To extend this point, he relies on a seemingly mundane example that turns out to be much more complex than it might appear: the creation of a website describing the mission of a research network. Of all the actors who might fall under the umbrella term "science communicator," he chooses to focus on the scientists themselves—those who actually created this statement. Their values can be found constantly intertwined with the scientific goals in this text.

Finally, Cynthia-Lou Coleman (chapter 14) articulates the imbalance that exists between the science that we think of as rational and free from bias and the perspectives of "others" that we think of as irrational and value-laden. She introduces the idea of biopolitics (that is, biology as politics) and that of validity claims as ways to gain a deeper understanding of the controversy that erupted over the 1996 discovery of an ancient skull found in the Columbia River and later referred to in news reports as "Kennewick Man." To briefly introduce this story, without giving away its somewhat ironic surprise ending, scientific interests claimed the skull was both "White" and theirs to analyze, while Native voices claimed the skull was from an ancestor and theirs to respectfully rebury.

These cases are all about science communication ethics issues as they arise in actual situations, which generally turn out to be quite complicated, even rather "messy" at times. The cases presented here were not chosen or designed to present a comprehensive or consistent view that might lead us to a broad general conclusion. If anything, they highlight a wide variety of "loose ends" and unresolved tensions across the range of cases, underscoring places where more work and additional reflection are needed.

However, there is a very important bottom line to this section: In studying the ethics of science communication, we should be mindful of the many boundaries the field is routinely asked to negotiate: between a reductionistic understanding of biology and the complexity of actual human choices; among citizen-consumers of genetic tests, the commercial enterprise that produces those tests, and the government that regulates their transactions; between the interests of local agricultural producers and those of mega-corporations marketing biotechnology; between values and norms and across disciplinary borders in the conceptualization of a research organization's mission; between the perspectives defined as "scientific" understandings of the world and what are often discounted as "nonscientific" alternatives.

Being an accomplished and ethical science communicator requires more than knowledge of the science itself or of ways to render it more readily understandable. It requires considerable social, political, and cultural awareness.

10

Just the Facts or Expert Opinion?
The Backtracking Approach to Socially
Responsible Science Communication

DANIEL J. MCKAUGHAN AND
KEVIN C. ELLIOTT

Science communicators often find themselves caught between two kinds of social responsibilities that sometimes stand in tension. On the one hand, society calls upon experts to discuss the significance of their work and to provide guidance on issues of social concern. We value expert opinion and we want to know what experts have to say. We invite their comments on all manner of issues upon which their work might be thought to bear. People with specialized cutting edge knowledge are often well-placed to participate in public discussions as responsible and well-informed citizens. Without channels by which they can speculate freely about potential ramifications of their work, a good many insights would be suppressed.

Yet on the other hand, we expect experts to be highly objective. Indeed, the degree to which the rest of us properly defer to them as experts often depends on public trust that what they say sticks closely to the facts. If experts manipulate the information that they communicate or if their comments smuggle in controversial judgments, even with the intention of benefiting the public, they open themselves to the criticism that they are undermining the process of democratic decision making by taking too much decision-making power away from the public. The dual goals of providing guidance and maintaining objectivity can be difficult to satisfy at once, and the challenge of balancing these conflicting responsibilities arises not just for experts but also, albeit in somewhat different ways, for science journalists and other communicators (McKaughan & Elliott, 2012, 2013).

We illustrate this tension using a case study of research on genetic and neurochemical changes that affect the social and sexual behavior of voles. As a response to this ethical challenge, we suggest that communicators should strive to promote "backtracking" on the part of information recipients. In other words, they should acknowledge significant value judgments that may have influenced their frames, interpretations, and assumptions, and they should try to clarify how one could arrive at alternative interpretations based on different value judgments.

The Ethical Challenge

As a society, we look to scientific experts as a source of at least two different kinds of information. First, we entrust specialists with the task of reliably reporting what has been established in a given area—public information established to a point that it is worthy of assent among a diverse and pluralistic society. Second, we often ask scientists to draw conclusions that require interpretive judgments or to summarize the potential ramifications of their work for socially relevant issues. For example, scientists have recently struggled to communicate effectively about issues like climate change, embryonic stem cell research, nuclear power and waste disposal, and evolutionary biology (see, e.g., Nisbet & Mooney, 2007; Pielke, 2007). It should not go unnoticed that in seeking to provide information that can guide societal decision making and shape public understandings of cutting-edge research, science communicators are regularly faced with a tension that, for short, we refer to as a tension between objectivity and guidance. We maintain that this tension can be regularly observed, and it gives rise to a number of difficult and very practical ethical questions. How much should scientists interpret their findings for policy makers? Is it appropriate for scientists to engage in advocacy or to make recommendations to policy makers? Should scientists allow their ethical or political values to influence their interpretations of ambiguous evidence? Should scientists adjust their standards of evidence when their claims could have major social ramifications?

These are significant ethical issues that fall under the rubric of what some have called the "ethics of expertise" (see, e.g., Douglas, 2008; Elliott, 2006; Hardwig, 1994; Resnik, 2001). Most of us, of course, value information that is both objective and easily understandable. We cannot give up either goal without significant social cost. Objectivity is important because both scientific and democratic principles call for scientists to avoid taking too much decision making power from the public, which can happen when they aggressively frame and interpret information (Elliott, 2011; Resnik, 2001; Resnik, 2009). If the presentation of scientific information becomes overly politicized, this can detract from the kind of neutrality on which some take the legitimate authority of experts to depend (Turner, 2001). But providing guidance is also important, for at least two reasons. First, efforts to be perfectly objective can end up backfiring by hiding values rather than eliminating them (Elliott & Resnik, 2014; Pielke, 2007). Second, efforts to be perfectly objective at the price of sacrificing guidance can result in confusion, poor interpretations of the evidence by the recipients of information, and ill-conceived decisions that harm society (Cranor, 1990; Elliott, 2011; Lehman-McKeeman & Kaminsky, 2013). In what follows we first illustrate this tension as it arises in a particular case and then go on to propose an approach to science communication that provides an alternative to attempts to resolve the tension by simply abandoning one or the other goal.

Moral Molecules and Love Drugs

Recently two structurally similar neuropeptides with a closely shared evolutionary history, oxytocin and vasopressin, have received attention owing to the roles that they play in mediating prosocial behavior. Oxytocin has been characterized as a "trust hormone," "cuddle chemical," "love drug," or even as "The Moral Molecule"—metaphors that seem to promise a panacea for many of our social ills. With respect to vasopressin, a gene (*V1aR*) associated with one of its receptor subtypes (the arginine vasopressin-1a receptor) has been heralded as a "gene for" "monogamy," "fidelity," or, when spinning it differently, "promiscuity," or "divorce." Attention-grabbing though they are, these beguiling catchphrases can frame discussion of larger issues of social and political interest or shape our understanding of ourselves as persons in ways that become as much sources of confusion as illumination (McKaughan, 2012; McKaughan & Elliott, 2012, 2013).

Much of our knowledge about these neuropeptides has come from research on voles, which are small rodents that resemble mice. For example, biologists have found intriguing connections between expression of the *V1aR* gene and sexual behavior among different species of voles. Specifically, prairie voles are relatively monogamous compared to montane and meadow voles. Prairie voles also have more vasopressin receptors in particular brain regions than montane and meadow voles because of a different promoter region for the vasopressin receptor gene (Young, Nilsen, Waymire, MacGregor, & Insel, 1999; Young, Lim, Gingrich, & Insel, 2001; Donaldson & Young, 2008; Insel, 2010). This genetic difference can be manipulated in various ways that result in striking changes in social and sexual behavior. For example, biologists have inserted the prairie vole gene into meadow voles and the modified meadow voles displayed partner preference (Lim et al., 2004). Moreover, by blocking receptors, biologists can eliminate partner preference in both prairie voles and the modified meadow voles (Winslow, Hastings, Carter, Harbaugh, &

Insel, 1993; Williams, Insel, Harbaugh, & Carter, 1994; Insel & Hulihan, 1995; Cho, DeVries, Williams, & Carter, 1999; Wang et al., 1999; Liu & Wang, 2003).

There are good biological reasons to be cautious about how similar changes might translate into other species, including humans (Fink, Excoffier, & Heckel, 2006; Goodson, Kelly, & Kingsbury, 2012; Young & Hammock, 2007). But there is also some intriguing evidence that suggests the neuropeptides vasopressin and oxytocin may be relevant to human social behaviors as well. A 2008 study led by Hasse Walum asked 552 couples who had been cohabitating for at least five years questions about their relationships and found that, in men, genetic variation in a microsatellite region of the human *V1aR* gene associates with the reported quality of their relationships. The authors conclude that

> the relatively small effect size of the *AVPR1A* polymorphism on traits tentatively reflecting pair-bonding in males observed in this study clearly does not mean that this polymorphism may serve as a predictor of human pair-bonding behavior on the individual level. However, by demonstrating a modest but significant influence of this gene on the studied behavior on the group level, we have provided support for the assumption that previous studies on the influence of the gene coding for *V1aR* on pair-bonding in voles are probably of relevance also for humans. (Walum et al., 2008, p. 14155)

A similar 2012 study of variants of the oxytocin receptor gene (*OXTR*) by this same research group found one single nucleotide polymorphism (SNP) of twelve examined—named rs7632287—that associates with traits reflecting pair-bonding in women (as measured by self-reports and questionnaires such as the Pair Bonding Scale) (Walum et al., 2012). Walum et al. 2012 understand the data as "possibly indicating that the well-described influence of oxytocin in voles could also be of importance to humans" (Walum et al., 2012, pp. 422, 419). Other studies of the influence of oxytocin in humans, some of which monitor the behavioral effects of oxytocin externally administered via nasal spray as subjects participate in various investment games, provide evidence for associations between oxytocin and prosocial behaviors such as trust (Kosfeld, Heinrichs, Zak, Fischbacher, & Fehr, 2005; Zak, 2005; Zak, Kurzban, & Matzner, 2005; Zak, Park, Ween, & Graham, 2006), generosity (Zak, Stanton, & Ahmadi, 2007), and reciprocity (Zak, 2011). There is also evidence that participants with a shorter variant in the length of the promoter sequence for the *V1aR* gene tend to behave less generously than players with the longer version (Knafo et al., 2008). So there is indeed a growing body of evidence that the neuropeptides oxytocin and vasopressin and their receptors can have an important influence on individual human social behaviors.

Such research invites big questions and, as one might imagine, scientists and journalists alike have found that it makes for exciting stories (McKaughan, 2012). One can find, often clustering around Valentine's Day, all manner of news stories about the discovery of a "cheating gene" and so on. We have identified several major frames that science communicators have been using in this case. One frame is "genetic determinism," according to which a particular gene or molecule controls even complex social behaviors such as sexual monogamy. According to a second frame, "humans are like voles," key aspects of human behavior are influenced by the same factors that are present in voles, so research findings in voles can be employed reliably for understanding humans. According to another frame, "saving your relationship," the lessons learned from the research can be used to develop drugs or other biotechnology innovations that can promote successful relationships or marriages.

Consider how each of these themes are woven together in a 2010 piece for *Psychology Today*. In "The Divorce Gene Explored: Should You Get Your Partner's DNA before Saying 'I Do'?" Dr. Shirah Vollmer, associate clinical professor of both psychiatry and family medicine at the David Geffen School of Medicine at UCLA, claims:

> This research opens the door to medication to treat infidelity. If we improve the reward of vasopressin, then we increase the likelihood of faithful marriages. It also changes the valence of fidelity. If infidelity is a genetic variant, should physicians treat it like hypertension or diabetes? On the other hand, perhaps the infidelity gene is closely linked to the charisma gene, and as such, it is part of the package of seduction.... Studies in prairie voles confirms [*sic*] my sense that we are all wired differently, and hence we come into the world with a different interface.... Perhaps we could sum it up this way: monogamy, one part family values, one part vasopressin responsiveness.

Vollmer's reference to a "divorce gene" evokes the "genetic determinism" frame, while her claim that "studies in prairie voles confirms my sense that we are all wired differently" expresses the "humans are like voles" frame, and her suggestion that physicians could treat infidelity like hypertension or diabetes evokes the "saving your relationship" frame.

Another frame, "dangers of social manipulation," highlights the potential for this research to open up worrisome developments that might have harmful social influences—perhaps that we are opening Pandora's Box or that there are dangers associated with technology development. In their 2009 book *Breeding Bio Insecurity: How U.S. Biodefense Is Exporting Fear, Globalizing Risk, and Making Us All Less Secure*, Lynn Klotz (a senior science fellow

at the Center for Arms Control and Non-proliferation) and Edward Sylvester
(professor at the Walter Cronkite School of Journalism and Mass Communication at Arizona State) cast the oxytocin research in rather alarmist terms:

> Spraying a captured city with a drug that would make its residents trust their
> new masters or make its insurgents throw roses at them may seem outlandish,
> and extremes in behavioral control are probably years off. But in 2005, Swiss
> researchers reported in the journal *Nature* that test subjects given a nasal spray
> of the natural brain hormone oxytocin showed markedly increased trust in
> those with whom they were dealing in social situations, compared with those
> given placebos. . . . The noted neuroscience researcher Antonio Damasio ob-
> served in an article accompanying the *Nature* research that sophisticated mar-
> keting and persuasive techniques already long used in sales and politics may
> actually work by stimulating the natural release of oxytocin in shoppers and
> voters. "Some may worry about the prospect that political operators will gen-
> erously spray the crowd with oxytocin at rallies of their candidates," he wrote.
> However, he concluded, "Civic alarm over such abuses should have started
> long before this study." Precisely. Indeed, warning signs have been present as
> the years have passed since the *Nature* publication, but they fall on deaf ears
> (Klotz and Sylvester, 2009, pp. 34–35).

Other writers emphasize the "understanding human nature" frame,
which suggests that this research may have implications that lead us to profoundly rethink our self-understanding. Setting out the thesis of his 2012
book on oxytocin, *The Moral Molecule: The Source of Love and Prosperity,* in
a video on his promotional website (www.moralmolecule.com), Paul Zak,
director of the Center for Neuroeconomics Studies at Claremont Graduate
University, explains:

> My book, *The Moral Molecule: The Source of Love and Prosperity,* details how
> I discovered a brain chemical, oxytocin, that makes us moral. It tells us why
> we can be so wonderful to others and sometimes also so cruel. . . . It tells us
> why we are who we are.

There are evident pros and cons to the framing of scientific information in
these sorts of narrative packages. Frames can, to be sure, help people begin to
make sense of a large and complex body of information or understand the potential significance of the voles research for a particular set of issues. But these
frames can also be very misleading and questionable, inviting public defer-
ence to experts on matters that may go well beyond what the evidence itself
shows. The frames also suppress additional information that renders the topics taken up much more complex and difficult to address with confidence
than they are made to seem. It is no part of our aim to discourage responsible discussion of these issues. Some of the scientists involved in this research

and philosophers who are well-informed about it have also thought that it has far-reaching implications. Leading vole researcher Larry Young takes the comparison between voles and humans on matters related to the vasopressin and oxytocin research quite seriously and has coauthored a popular book with Brian Alexander, *The Chemistry between Us: Love, Sex, and the Science of Attraction* (2012). Clearly there are interesting questions here, and we need people who are well-informed to be thinking about the significance of cutting edge research. (For a scientifically informed discussion of ethical issues involved in the development of potential "love drugs" we refer readers to Earp, Sandberg, and Savulescu, 2015.) Our interest in the present essay is in whether there are some fairly simple and straightforward steps that science communicators could take to elevate the current level of public discussion and to preserve their objectivity while at the same time promoting understanding.

Backtracking

The frames at work in the examples from the previous section illustrate how scientific research is taken up in discussion of issues of broad social interest. As we emphasized above, we are interested in how science communicators might present information in a way that does not preclude them from sharing their reflections on the meaning and significance of their work, while doing so in a way that respects the self-determination of their readers and listeners (Elliott, 2010). In other words, we want communicators to equip information recipients who are not in as good a position to evaluate the evidence as experts, who may approach the interpretive process with different assumptions and values, to begin to think through the issues responsibly for themselves.

How might we navigate between the Scylla of sacrificing guidance and the Charybdis of compromising objectivity? The course we suggest is that those involved in the process of science communication—including scientists, journalists, and their critics—take a more active role in the effort to make key values and assumptions visible (Elliott & Resnik, 2014). If experts were discouraged from freely sharing their perspectives on matters of controversy or on issues where the evidence is ambiguous, we would be deprived of many valuable insights. Yet, public trust in the reliability of science is predicated on the assumption that experts are not simply passing off controversial personal value judgments as well-established scientific conclusions. We have introduced the term "backtracking" as a way of describing this process of clarifying the ways that value judgments enter into scientific research and communication (McKaughan & Elliott, 2013).

To backtrack is to go back over a path, to return to a previous point. In the context of science communication, the "path" is a metaphor for the value judgments that result in particular interpretations or frames. In this context, we use the term "value judgment" as it is typically used in the philosophy of science literature. In other words, it refers not only to ethical or political decisions but also to any judgment that requires weighing various forms of evidence or other desiderata. For example, philosophers have argued that value judgments are involved in setting standards of evidence (Douglas, 2009), choosing terminology (Elliott, 2009), making assumptions about how to interpret evidence (Longino, 1990), and framing scientific questions or problems (Anderson, 2004; McKaughan & Elliott, 2013). Experts enable information recipients to backtrack by making them more aware of the interpretive "path" that has been taken, i.e., by highlighting the major value judgments involved in reaching a particular conclusion or frame. This goal can sometimes be advanced by clarifying the weaknesses of particular assumptions, by qualifying one's conclusions, or by acknowledging possible alternative interpretations of the available evidence. The goal of these efforts is to ease the tension between communicating in a manner that guides the public while maintaining objectivity.

Consider how backtracking might play out in practice. Suppose that a scientist were disseminating information that was debated or subject to multiple interpretations in the scientific community. An initial step would be for her to acknowledge the range of different interpretations available when writing journal articles or speaking with reporters or writing for a popular audience. Another important step would be for her to clarify which interpretations were best supported by the available evidence and which ones were more speculative. Because individual scientists are subject to a range of idiosyncratic preferences and biases (Elliott & Resnik, 2015), it would be particularly valuable for her to clarify the views of the scientific community as a whole about the evidence for different interpretations rather than focusing solely on her own views. Depending on the context and the amount of time available for her to discuss her work, she could also clarify the major assumptions or value judgments that explain why some scientists arrive at different interpretations than others. Finally, she could point interested people to other sources for obtaining further information. Ideally, journalists who work with scientists to write popular stories could help them to take these steps.

Unfortunately, individual scientists may not always be consciously aware of the ways in which they steer their listeners toward some interpretations rather than others. Therefore, in order to facilitate backtracking, it is important for scientists to take additional steps to assist their readers and listeners.

One important step is to disclose significant interests or values that could (consciously or subconsciously) affect their interpretations (Elliott & Resnik, 2014). For example, many journals have begun to require that scientists disclose the funding sources for their research, and it would be valuable for scientists to provide this information whenever they discuss their work. Another step that individual scientists can take is to make their data publicly available after publication so that others can more effectively evaluate their analyses and interpretations (Soranno, Cheruvelil, Elliott, & Montgomery, 2015). Finally, the scientific community and broader society can take additional steps to help promote backtracking about important scientific issues. One approach is to create deliberative forums in which multiple scientists and concerned citizens can discuss the status of important scientific issues (see, e.g., Elliott, 2011). Scholars and citizens can also work with NGOs and civil society groups to help maintain robust conversations about scientific issues that are significant for society.

Our backtracking framework can be made even more concrete by considering how it might apply in the case of the oxytocin and vasopressin research. We think that three lessons may be helpful in this and other cases: (1) a number of relatively simple caveats and clarifications can be employed; (2) some frames or interpretations are so likely to be misinterpreted that they should perhaps be minimized or avoided; and (3) the entire community involved in the process of science communication—including experts, journalists, information recipients, and various critics—should take responsibility for the roles they can play in promoting the process of backtracking.

First, consider how a few cautionary comments can assist backtracking in the voles case. One point that can easily be confused when employing the "humans are like voles" frame is that the biologists' technical term "monogamy" can differ substantially from ordinary uses of monogamy, where people typically have in mind a sexually exclusive relationship with one partner. Biologists typically are talking about social monogamy rather than sexual exclusivity (i.e., genetic monogamy). Social monogamy involves preferential association and cohabitation with a partner but is compatible with extra-pair copulations that would be incompatible with "monogamy" or "fidelity" as these are typically used in ordinary discourse for describing long-term human relationships (Young & Hammock, 2007). Making this basic point need not be too complex but is simply a matter of noting a few caveats, and can go a long way toward dispelling misconceptions that might arise in connection with talk about genes for divorce or medications that might treat infidelity.

A second point worth emphasizing in journalistic presentations exploring the "humans are like voles" frame is to make the simple point that what we are learning about the voles may not translate into human behavior in any clear or straightforward way. It is worth noting that scientists are often already trying to make these clarifications. For example, Thomas Insel, one of the key early figures in this work and past director of the National Institute of Mental Health, recommends that we take oxytocin and vasopressin receptor genes as "reasonable candidates to study in humans, recognizing that species differences are the hallmark of nonapeptide evolution" (Insel, 2010). Such comments help not only to make clear why the frame is suggestive and scientifically interesting but also to caution against jumping to conclusions about how things might work in humans and alert us to important differences between species.

Similar points can be made about the "genetic determinism" frame, for example, by reminding readers that developmental factors and social experience are also extremely important in behavior not only in humans but also in voles. As McGraw et al. note, while many of the mechanisms influencing behavior are not yet well understood, we do know that "the prairie vole brain is exquisitely sensitive to the influence of social experience which shapes the expression of behaviorally relevant genes" (McGraw, Thomas, & Young, 2008, p. 1). There are many other ways that experts can clarify the issues, for example by pointing out that "genes for" phenotypic traits are frequently not all-or-nothing and depend heavily on environmental context. Again, these are relatively simple points that experts can often make quickly and on the fly even during a live interview.

A second lesson in this and other cases is that some frames or interpretations should be minimized or avoided. Not all frames are equally useful or well grounded, and some depend on particularly controversial or questionable value judgments. Given the difficulties associated with undoing the effects of framing, sometimes communicators might decide that a particular frame is just so questionable, liable to be misconstrued, or socially problematic that it should be avoided. For example, reporting a story on "The cheatin' gene: Researchers find men may be genetically predisposed to cheat" for the NBC Nightly News, anchor Brian Williams tells us:

> Throughout history men have come up with all sorts of excuses for behaving badly. Now it appears they have a new one. It's in their genes, apparently. This is a line of research that started with rodents called voles. Now it's being applied to humans. Our chief science correspondent Robert Bazell explains.

Rightly or wrongly, many of us place a high degree of trust in the nightly news. But what is the likely uptake of framing the science in this way? After hearing this kind of information, a woman who was interviewed by NBC's

Today Show stated: "I would want to have my mate tested. . . . And I am single and that would secure my marriage." If this is the kind of message that one will get across to society—an impression of scientific consensus that the complex behaviors in human relationships are in some straightforward way controlled by our genes and that it is now sensible to think about making major life decisions on this basis—science communicators should think twice about whether framing the findings in this way is really the best way to provide guidance about the state of current science.

It is of course true that an individual scientist who is being interviewed by a journalist writing a story on their work might have very little control over how the research is eventually framed. But this leads us to our third point, namely, that backtracking, as we are envisioning it, is a responsibility of the entire community of those involved in science communication—including experts, members of the media, information recipients, and critics. We need journalists to make clarifications, to distinguish claims that are better supported from those that are more speculative, and to interview enough figures to get an understanding of the range of perspectives in the scientific community and the value judgments responsible for important disagreements. Other scholars and critics can also challenge journalists and scientists who fail to promote backtracking. Finally, if an expert feels uncomfortable with the direction a reporter seems to be taking a story, it may not be difficult to say so and to suggest an alternative way of seeing things. For example, rather than focusing on marriage and sex, one could explore the possibilities such research might open up for treating autism spectrum disorders or encourage people to think about other aspects of the research's significance. Alternatively, an expert could insist, if they are to be quoted, that certain qualifications be included in the piece.

Potential Objections

In order to clarify the strengths and weaknesses of the backtracking approach to socially responsible science communication, it may be helpful to address several potential objections. First, it might seem that efforts to promote backtracking clash with one of the two fundamental ethical goals or norms that we presented at the beginning of this paper. For example, we noted that scientists have not only the goal of maintaining objectivity but also the goal of providing guidance to information recipients. One might worry that if backtracking involves acknowledging complications and alternative interpretations of the available evidence in an effort to promote objectivity, it complicates communication in a manner that can create confusion and sacrifice effective guidance. Our response is that while backtracking may complicate communication efforts somewhat, it is still important to pursue as a way of balancing the dual goals of objectivity and guidance. If communicators take the liberty of aggressively interpreting and framing scientific work without putting efforts into acknowledging weaknesses and alternatives, they arguably move too far in the direction of sacrificing objectivity.

Alternatively, one might think that backtracking clashes with the goal of promoting objectivity, given that it allows for interpretive leeway when communicating scientific information. One might also suppose that the backtracking approach is overly influenced by the notion that scientists should interpret information for the public as voices of authority. In contrast, one might argue that scientists should merely present the range of available evidence in as perspicuous a fashion as possible so that information recipients can interpret this information for themselves. But it is unrealistic to think that efforts to present evidence responsibly can alleviate the need for backtracking. First, a great deal of literature indicates that efforts to purify scientific conclusions of all value judgments are likely to be counterproductive insofar as they merely suppress attention to important judgments that remain unavoidable (e.g., Longino, 1990; Douglas, 2009; Elliott & Resnik, 2014). Second, in an effort to "stick to the facts" and avoid any interpretive judgments, scientists run the risk of harming society by leaving information recipients confused about how to understand the available evidence (Cranor, 1990, p. 139; Elliott, 2011, p. 67). Thus, our backtracking approach remains an important compromise between the dual ethical goals of promoting objectivity and providing guidance.

A second worry is that our call for backtracking might seem to rest on a false assumption—namely, that there are "neutral," value-free scientific facts available. If one adopts the more realistic assumption that data are theory-laden, one might worry that scientists cannot backtrack to any universally accepted starting points. Our response to this worry is that our concept of backtracking does not presuppose the existence of neutral, universally accepted scientific facts. Rather, it merely requires that some frames or interpretations incorporate more controversial value judgments than others. Backtracking can involve flagging this point or otherwise giving the recipients of information a better sense of the interpretive "lay of the land."

Third, one might worry that we are asking too much of the scientific community. This concern might take at least two forms. First, it might seem that scientists do not have enough time to add these concerns about scientific communication to all the other demands on their time and attention in the current ultra-competitive scientific environment (Alberts, Kirschner,

Tilghman, & Varmus, 2014). Second, it might seem unlikely that scientists can recognize all of the significant assumptions, interests, and values that affect their work. Our response to these worries is that one does not need to put all the pressure on scientists to achieve effective backtracking. They can be part of broader networks of people and institutions—including NGOs, activist groups, interdisciplinary communities of scholars, and journalists—that work together to identify important value judgments and explore ways of acknowledging them.

A fourth concern is that our goal of promoting backtracking may be psychologically unrealistic. One might worry that once people are influenced by a particular frame, the influences of the value judgments that support that frame cannot be completely undone. For example, Brendon Larson has noted that even when the weaknesses of particular metaphors are pointed out to the public, this further education sometimes does little to correct the misunderstandings caused by them (Larson, 2011, p. 210). We do not think that this difficulty completely negates the importance of trying to promote backtracking, but it does highlight the importance of reflecting carefully about what value judgments should guide the initial frames or interpretations that are communicated. Researchers should not blithely advocate questionable interpretations and assume that they can easily enable information recipients to backtrack away from the initial interpretations.

It may be fruitful to compare our call for careful reflection about the choice of initial frames to the recommendations of Richard Thaler and Cass Sunstein in their book *Nudge* (2008). They point to literature in cognitive psychology indicating that our choices are influenced to a surprising extent by situational factors. For example, while we move freely through a cafeteria or a grocery store choosing items to purchase, psychological research indicates that our choices are also significantly influenced by the locations in which items are placed. Thaler and Sunstein argue that the best response to these psychological findings is to adopt a philosophical approach that they call libertarian paternalism. According to this approach, those in a position to influence people's behavior should steer people in ways that benefit them, but they should also seek to provide opportunities for people to "opt out" of those behaviors if they want to. Similarly, we suggest that science communicators should use good judgment in basing their frames or interpretations on value judgments that are socially beneficial, but they should do their best to help information recipients shift to other frames or interpretations if they choose to do so.

A fifth worry about our suggestion to promote backtracking is that the characteristics of mass media may make it exceedingly difficult to follow our recommendations. For example, media outlets are typically under pressure to make money by generating public interest in their news reports. As a result, journalists are often pressured to frame their stories in the ways that are most engaging and perhaps even sensational, not in ways that promote backtracking. Moreover, because print media outlets are struggling financially, they are often not able to afford dedicated science writers. This makes it even more difficult for journalists to develop the expertise to present information in a nuanced manner that promotes backtracking. Our response to this worry is fairly simple: something is better than nothing. Even if the nuances that scientists try to provide are subsequently garbled as they pass through the media, it is better to try to provide ethically important nuances than to do nothing at all. Furthermore, there are sometimes relatively easy steps that scientists or journalists can take in order to promote backtracking. For example, at the end of a relatively simple news story they can provide web links so that those who want more detailed information can explore an issue further.

Conclusion

In the German fairy tale "Hansel and Gretel," a brother and sister leave a trail of stones or bread crumbs to find their way back out of the forest. Similarly, we suggest, even offering just a few crumbs or hints that clarify and acknowledge key value judgments that influence the interpretive process can helpfully serve the public and at the same time serve science by building public trust in the communicative process. Steps that enable backtracking need not be onerous or terribly impractical. We are not asking scientists to do the impossible. Even just a few well-chosen caveats can sometimes substantially elevate the level of public discussion. To be sure, just as sometimes bread crumbs get eaten by birds, the various caveats, clues, or signposts left by communicators are sometimes lost in the flood and shuffle of information. But the fact that we sometimes, even typically, fail to perfectly realize the goal of enabling backtracking does not imply that it cannot still serve as a useful norm toward which to strive. There is always a danger that people get lost in the woods, but scientists and journalists can help people to identify the path they are on and to find alternative paths by clarifying the key value judgments involved in their interpretive work.

We have emphasized the importance of transparency in science communication. The public has much to gain from science communicators who are willing to weigh in on issues of public interest. But such comments are of greater worth when the values that underlie them are visible in the light of day so that these too can be more readily subjected to public scrutiny and to open and ongoing discussion. Information geared at enabling backtracking helps non-experts to gauge the extent to which information could be interpreted differently or whether individuals might reasonably come to different conclusions about a particular set of claims. By enabling backtracking, science communicators help to "thread the needle" between achieving objectivity and providing guidance to society.

References

Alberts, B., Kirschner, M., Tilghman, S., & Varmus, H. (2014). Rescuing US biomedical research from its systemic flaws. *Proceedings of the National Academy of Sciences 111*(16), 5773–5777.

Anderson, E. (2004). Uses of value judgments in science: A general argument, with lessons from a case study of feminist research on divorce. *Hypatia, 19*, 1–24.

Bazell, R. (2008, September 2). The cheatin' gene: Researchers find men may be genetically predisposed to cheat. *NBC Nightly News* with Brain Williams. Retrieved from http://www.msnbc.msn.com/id/3032619/#26512696.

Cho, M. M., DeVries, A. C., Williams, J. R., & Carter, C. S. (1999). The effects of oxytocin and vasopressin on partner preferences in male and female prairie voles (*Microtus ochrogaster*). *Behavioral Neuroscience 113*(5), 1071–1079.

Cranor, C. (1990). Some moral issues in risk assessment. *Ethics, 101*, 123–143.

Donaldson, Z. R., & Young, L. J. (2008). Oxytocin, vasopressin, and the neurogenetics of sociality. *Science, 322*(5903), 900–904.

Douglas, H. (2008). The role of values in expert reasoning. *Public Affairs Quarterly, 22*, 1–18.

Douglas, H. (2009). *Science, policy, and the value-free ideal*. Pittsburgh, PA: University of Pittsburgh Press.

Earp, B. D., Sandberg, A., & Savulescu, J. (2015). The medicalization of love. *Cambridge Quarterly of Healthcare Ethics, 24*(3), 323–336.

Elliott, K. C. (2006). An ethics of expertise based on informed consent. *Science and Engineering Ethics, 12*, 637–661.

Elliott, K. C. (2009). The ethical significance of language in the environmental sciences: Case studies from pollution research. *Ethics, Place, and Environment, 12*, 157–173.

Elliott, K. C. (2010). Hydrogen fuel-cell vehicles, energy policy, and the ethics of expertise. *Journal of Applied Philosophy, 27*, 376–393.

Elliott, K. C. (2011). *Is a little pollution good for you? Incorporating societal values in environmental research*. New York, NY: Oxford University Press.

Elliott, K. C., & Resnik, D. B. (2014). Science, policy, and the transparency of values. *Environmental Health Perspectives, 122*, 647–650.

Elliott, K. C., & Resnik, D. B. (2015). Scientific reproducibility, human error, and public policy. *BioScience, 65*, 5–6.

Fink, S., Excoffier, L., & Heckel, G. (2006). Mammalian monogamy is not controlled by a single gene. *Proceedings of the National Academy of Sciences of the United States, 103*, 10956–10960.

Goodson, J. L., Kelly, A. M., & Kingsbury, M. A. (2012). Evolving nonapeptide mechanisms of gregariousness and social diversity in birds. *Hormones and Behavior, 61*, 239–250.

Hardwig, J. (1994). Toward an ethics of expertise. In D. Wueste (Ed.), *Professional ethics and social responsibility* (pp. 83–102). Lanham, MD: Rowman and Littlefield.

Insel, T. R. (2010). The challenge of translation in social neuroscience: a review of oxytocin, vasopressin, and affiliative behavior. *Neuron, 65,* 768–779.

Insel, T. R., & Hulihan, T. J. (1995). A gender-specific mechanism for pair bonding: Oxytocin and partner preference formation in monogamous voles. *Behavioral Neuroscience, 109*(4), 782–789.

Klotz, L., & Sylvester, E. 2009. *Breeding bio insecurity: How U.S. biodefense is exporting fear, globalizing risk, and making us all less secure.* Chicago, IL: University of Chicago Press.

Knafo, A., Israel, S., Darvasi, A., Bachner-Melman, R., Uzefovsky, F., Cohen, L., . . . & Ebstein, R. P. (2008). Individual differences in allocation of funds in the dictator game associated with length of the arginine vasopressin 1a receptor RS3 promoter region and correlation between RS3 length and hippocampal mRNA. *Genes, Brain, and Behavior, 7,* 266–275.

Kosfeld, M., Heinrichs, M., Zak, P. J., Fischbacher, U., & Fehr, E. (2005). Oxytocin increases trust in humans. *Nature, 435,* 673–676.

Larson, B. (2011). *Metaphors for environmental sustainability: Redefining our relationship with nature.* New Haven, CT: Yale University Press.

Lehman-McKeeman, L., & Kaminsky, N. (2013). The hazards of playing it safe: Perspectives on how the Society of Toxicology should contribute to discussions on timely issues of human and environmental safety. *Toxicological Sciences, 136,* 1–3.

Lim, M. M., Wang, Z., Olaza'bal, D. E., Ren, X., Terwilliger, E. F., & Young, L. J. (2004). Enhanced partner preference in a promiscuous species by manipulating the expression of a single gene. *Nature, 429,*754–757.

Liu, Y., & Wang, Z. X. (2003). Nucleus accumbens oxytocin and dopamine interact to regulate pair bond formation in female prairie voles. *Neuroscience, 121*(3), 537–544.

Longino, H. (1990). *Science as social knowledge.* Princeton, NJ: Princeton University Press.

McGraw, L. A., Thomas, J. W., & Young, L. J. (2008). White paper proposal for sequencing the genome of the prairie vole (*Microtus ochrogaster*). Sequencing proposal submitted to the National Human Genome Research Institute (NHGRI) by research groups at Emory and accompanied by letters of support by leading scientists who work with voles, 1–64. Retrieved from http://www.genome.gov/Pages/Research/Sequencing/SeqProposals/VoleWhitePaper_and_LOS.pdf

McKaughan, D. J. (2012). Voles, vasopressin, and infidelity: A molecular basis for monogamy, a platform for ethics, and more? *Biology and Philosophy, 27*(4), 521–543.

McKaughan, D. J., & Elliott, K. C. (2012). Voles, vasopressin, and the ethics of framing. *Science, 338*(6112), 1285.

McKaughan, D. J., & Elliott, K. C. (2013). Backtracking and the ethics of framing: Lessons from voles and vasopressin. *Accountability in Research, 20*(3), 206–226.

Nisbet, M. C., & Mooney, C. (2007). Framing science. *Science, 316,* 56.

Pielke, R. (2007). *The honest broker: Making sense of science in policy and politics.* Cambridge, UK: Cambridge University Press.

Resnik, D. (2001). Ethical dilemmas in communicating medical information to the public. *Health Policy, 55,* 129–149.

Resnik, D. (2009). *Playing politics with science: Balancing scientific independene and government oversight.* New York, NY: Oxford University Press.

Soranno, P., Cheruvelil, K., Elliott, K., & Montgomery, G. (2015). It's good to share: Why environmental scientists' ethics are out of date. *BioScience, 65,* 69–73.

Special Edition: His cheatin' genes? New science links biology, monogamy. (2008, September 3). NBC's *Today Show*. Retrieved from http://today.msnbc.msn.com/id/26523972/#.UXMgB16D9R0

Thaler, R. H., & Sunstein, C. R. (2008). *Nudge: Improving decisions about health, wealth, and happiness.* New Haven, CT: Yale University Press.

Turner, S. (2001). What is the problem with experts? *Social Studies of Science, 31,* 123–149.

Vollmer, S. (2010, January 3). The divorce gene explored: Should you get your partner's DNA before saying "I Do"? *Psychology Today.* Retrieved from http://www.psychologytoday.com/blog/learning-play/201001/2010-the-divorce-gene-explored

Walum, H., Westberg, L., Henningsson, S., Neiderhiser, J. M., Reiss, D., Igl, W., . . . Lichtenstein, P. (2008). Genetic variation in the vasopressin receptor 1a gene (AVPR1A) associates with pair-bonding behavior in humans. *Proceedings of the National Academy of Sciences of the United States, 105,* 14153–14156.

Walum, H., Lichtenstein, P., Neiderhiser, J. M., Reiss, D., Ganiban, J. M., Spotts, E. L., . . . Westberg, L. (2012). Variation in the oxytocin receptor gene is associated with pair-bonding and social behavior. *Biological Psychiatry 71:* 419–426.

Wang, Z. X., Yu, G. Z., Cascio, C., Liu, Y., Gingrich, B., & Insel, T.R. (1999). Dopamine D2 receptor-mediated regulation of partner preferences in female prairie voles: a mechanism for pair bonding. *Behavioral Neuroscience, 113,* 602–611.

Williams, J. R., Insel, T. R., Harbaugh, C. R., & Carter, C. S. (1994). Oxytocin administered centrally facilitates formation of a partner preference in female prairie voles (*Microtus ochrogaster*). *Journal of Neuroendocrinology, 6,* 247–250.

Winslow, J. T., Hastings, N., Carter, C. S., Harbaugh, C. R., & Insel, T. R. (1993). A role for central vasopressin in pair bonding in monogamous prairie voles. *Nature, 365,* 545–548.

Young, L. J., & Alexander, B. (2012). *The chemistry between us: Love, sex, and the science of attraction.* New York, NY: Penguin.

Young, L. J., & Hammock, E. A. D. (2007). On switches and knobs, microsatellites and monogamy. *Trends in Genetics, 23*(5), 209–212.

Young, L. J., Lim, M. M., Gingrich, B., & Insel, T. R. (2001). Cellular mechanisms of social attachment. *Hormones and Behavior, 40,* 133–138.

Young, L. J., Nilsen, R., Waymire, K. G., MacGregor, G. R., & Insel, T. R. (1999). Increased affiliative response to vasopressin in mice expressing the V1a receptor from a monogamous vole. *Nature, 400,* 766–768.

Zak, P. J. (2005). Trust: A temporary human attachment facilitated by oxytocin. *Behavioral and Brain Sciences, 28,* 368–369.

Zak, P. J. (2011). Moral markets. *Journal of Economic Behavior & Organization, 77*(2): 212–233.

Zak, P. J. (2012). *The moral molecule: The source of love and prosperity.* New York, NY: Dutton/Penguin Press.

Zak, P. J., Kurzban, R., & Matzner, W. T. (2005). Oxytocin is associated with human trustworthiness. *Hormones and Behavior, 48,* 522–527.

Zak, P. J., Park, J., Ween, J., & Graham, S. (2006). An fMRI study of interpersonal trust with exogenous oxytocin infusion. *Society for Neuroscience,* Program No. 2006-A-130719-SfN.

Zak, P. J., Stanton, A. A., & Ahmadi, S. (2007). Oxytocin increases generosity in humans. *PLOS ONE, 2*(11), e1128.

11

Controversy, Commonplaces, and Ethical Science Communication: The Case of Consumer Genetic Testing

LORA ARDUSER

Genetic testing marketed directly to consumers through television, print advertisements, and the internet has been gaining prominence over the past several years (Javitt, Stanley, & Hudson, 2004). In fact, the global direct-to-consumer (DTC) genetic testing market is projected to exceed $230 million by 2018 (Global Industry Analysts, 2012). Two of the most widely known DTC genetic testing companies in the United States met their own milestones in recent years. In 2014, the Provo, Utah–based company Ancestry.com surpassed 500,000 genotyped members. In 2015, 23andMe announced that it had genotyped its one-millionth customer. Proponents argue that such testing allows people to be more proactive in their health care, but the US Food and Drug Administration (FDA) has concerns about people being misled by the results of DTC genetic testing and making important decisions about treatment or prevention based on inaccurate, incomplete, or misunderstood information.

In 2006 when DTC genetic testing first came onto the market, regulation and oversight of the products were incomplete and confusing at best. The Federal Trade Commission (FTC), the FDA, and the Centers for Disease Control and Prevention (CDC) published documents warning consumers to be cautious of claims made by such companies and to be sure to include a medical professional in interpreting genetic test results (Vorhaus, 2010). This attention to the importance of the interpretation of DTC genetic test results is of particular interest to interdisciplinary questions about ethical science communication because the multiple stakeholders involved in DTC genetic testing communication often have disparate interests. People in the public may be interested in self-discovery, as ABC news reporter Dan Spindle describes in the beginning of his 2013 report on the controversy between the

DTC company 23andMe and the FDA: "For just $99 and about five minutes online, I started my journey of self-discovery" (Spindle, 2013, n.p.). The FDA's mission is to promote and protect public health. As a private corporation, 23andMe is interested in profit and, according to its website, the advancement of science.

These competing interests raise the question of how DTC test results can be ethically communicated when the involved parties define a "good end" in different ways. This chapter addresses this question through an examination of texts produced by the stakeholders in the 23andMe controversy. The goal of the analysis is to devise a framework that describes what ethical communication could look like in this situation. Such a framework has implications for how genetic testing policy is developed, how test kits are marketed, and how test results are communicated. It also more generally informs ways contested values can be negotiated in participatory science.

Communication Concerns about DTC Genetic Testing

Genetic testing has moved from being an activity practiced exclusively in a laboratory to one that can be undertaken by people in their homes buying saliva test kits, spitting into a vial, and mailing the kit back to a lab to get the results. Celeste Condit in her 1999 book, *The Meanings of the Gene: Public Debates about Human Heredity*, raised questions about ethics and communication that only multiply in this new environment. These concerns include issues of trust and the credibility of the companies that sell these tests (Delfanti, 2011), difficulties with national regulatory mechanisms for tests marketed and sold online (William-Jones, 2006), and DTC companies' ability to meet the standards of analytic validity, clinical validity, and clinical utility that traditional testing labs meet (Curnutte & Testa, 2012).

The public's reception of DTC genetic-testing messages (Einsiedel & Geransar, 2009) is also important. This reception of these messages has been explored in terms of whether or not people would act upon the information they received from online genetic testing sources and issues of trust and literacy. For example, Critchley, Nicol, Otlowski, and Chalmers (2015) found in their telephone survey of 1,000 Australians that significantly less of the sample was likely to order a DTC genetic test administered by a company as compared to one administered by a medical professional, citing such reasons as their lack of trust in the company being able to protect their privacy as well as the inability of the company to provide them with genetic counseling services. Similarly, in examining attitudes toward DTC testing for the breast cancer susceptibility gene (BRCA), Perez et al. (2011) found that among the

84 women they interviewed at a high-risk clinic, the majority of the women held favorable attitudes toward DTC testing for BRCA but did not support online testing.

Along with issues of trust in sources, the public reception of DTC results is intertwined with questions of genetic literacy (Pearson & Liu-Thompkis, 2012), understanding, and expertise. Serious concerns have been expressed as to whether or not consumers possess the knowledge to make sound decisions about the use of the information from DTC genetic testing (Pearson & Liu-Thompkins, 2012). To help assuage this concern, scientists and medical professionals in general have deployed a striking range of metaphors, a prevalent and important vehicle of public communication, to communicate their science and promote its value (Nelkin, 2001). Much of the literature about communicating genetic information focuses on the use of these devices. Metaphors in genetic science have included that of the blueprint (Condit, 1999). Wilson (2002) also describes text metaphors used in relation to the Human Genome Project, where letters are bases, words are genes, and the complete genome is the book. Others have detailed the metaphors of maps and cartography (Lippman, 1992; Rothman, 1998). More recently metaphors of maps, blueprints, codes, and texts have been critiqued as inherently limited because they misrepresent or oversimplify science (Lopez, 2007). As an alternative, in his analysis of the traveling science exhibition titled The Geee! in Genome, Lopez offers a musical metaphor of nucleotide bases as notes that produce the "music of life" (p. 9), a metaphor that can better capture what Lopez calls "our wider cultural schemas, which frame individuals as being unique and their lives as journeys of self-discovery" (2007, p. 27). Asserting that the scientific community and the lay public use different genetic metaphors, Gronnvoll and Landau (2010) likewise propose new genetic metaphors, suggesting "dance" and "band."

The 23andMe Controversy

Research about 23andMe specifically has focused on the company's dual mission of generating profit and advancing science (Harris, Wyatt, & Kelly, 2013). In 2006, when 23andMe opened for business, the company positioned its product as a "fun way to learn a little genetics using yourself as a test subject" (Seife, 2013). The goal, the company stated on its website in 2007, "is to connect you to the 23 paired volumes of your own genetic blueprint (plus your mitochondrial DNA), bringing you personal insight into ancestry, genealogy, and inherited traits." As time passed, 23andMe put less emphasis on the ancestral component of its service and began to emphasize marketing their services as a way of "predicting and even preventing health problems" (Seife, 2013).

Although the FDA was still defining its role in the regulation of companies like 23andMe, as more companies began to offer DTC genetic testing products and more consumers began to purchase them, the agency became increasingly involved in regulating what they defined as a medical device.[1] As medical devices, the genetic test kits require FDA review and approval before they can be marketed. In what could be defined as an effort to catch up with the rapid activities taking place in the field of DTC genetic testing, in 2010 the agency issued warning letters to 17 companies. The letters stated that the FDA wanted to meet with company representatives to discuss whether the test kits they were advertising and selling required review by the FDA. 23andMe received its warning letter on November 22, 2013. The letter directed the company to immediately discontinue marketing the 23andMe Saliva Collection Kit and Personal Genome Service (PGS) until it received marketing authorization for the device from the FDA (FDA, 2013).

It is clear in the FDA's warning letter to 23andMe that actions taken related to diagnosis, prevention, or treatment of disease are defined as being the result of communication. At the end of the third paragraph of the letter, Alberto Gutierrez, director of the Office of In Vitro Diagnostics writing on behalf of the FDA, states that "The risk of serious injury or death is known to be high when patients are either non-compliant or not properly dosed; combined with the risk that a direct-to-consumer test result may be used by a patient to self-manage, serious concerns are raised *if test results are not adequately understood* by patients or *if incorrect test results are reported*" (FDA, 2013, n. p., emphasis added). This sentence indicates that people using the kits are at risk because of ineffective communication (i.e., communication that includes inaccurate information or communication that may be accurate but misunderstood by the audience).

Such language is based on two assumptions. The first is a view of communication that conceptualizes the public as having knowledge deficits about technoscientific issues (Felt, Fochler, Müller, & Strassnig, 2009). The second

---

1. According to the FDA a medical device is "an instrument, apparatus, implement, machine, contrivance, implant, in vitro reagent, or other similar or related article, including a component part, or accessory that is either 1) recognized in the official National Formulary or the United States Pharmacopoeia, 2) used in the diagnosis of disease or in the cure, treatment, or prevention of disease, or 3) intended to affect the structure or any function of the body of man or other animals, and which does not achieve its primary intended purposes through chemical action within or on the body of man or other animals and which is not dependent upon being metabolized for the achievement of any of its primary intended purposes" (FDA, 2014, n.p.).

is an assumption that involves prioritizing the knowledge and authority of science and medicine over both business and the general public. Both of these assumptions get in the way of an ethical communicative framework that considers all stakeholder values, and, in turn, values all stakeholders' knowledge and experience.

Furthermore, because communication in this situation has to cross lines that often instill distrust and assumptions are made about stakeholders that may or may not be true (i.e., corporations are only interested in profits and the government cannot get anything done; the general public is unknowledgeable about technoscientific subjects), science communication that relies on a typical "top down" process is problematic for creating an ethical communication framework. To uncover the contested values and expectations between multiple stakeholders in DTC communication requires a pluralistic approach that first identifies the topics of concern (Ross, 2013; Walsh, 2010) or "commonplaces" (McKeon, 1987) of each stakeholder. Such commonplaces are based on a system of special and general topics devised by Aristotle for teaching students to be persuasive speakers. However, they are applicable in many situations of public discourse (Miller & Selzer, 1985), including those of science-related argumentation (Ross, 2013), communicating in complicated situations such as climate change debates and research publications in scientific, technical, mathematical, and engineering (STEM) fields (Walsh, 2010), and communicating about genetic testing public policy (Murphy & Maynard, 2000).

Identifying and Organizing the Controversy's Commonplaces

To identify these topics of mutual concern, this chapter presents an analysis of texts from the stakeholders in the 23andMe controversy: documents authored by the FDA, 23andMe, and customers of 23andMe. Documents from the FDA and 23andMe include the FDA's warning letter to 23andMe; the text of the March 8–9, 2011, hearing; and the 23andMe website. To acquire an understanding of the interests and concerns of customers, 73 customer reviews posted on Amazon.com between 2010 and 2013 were included in the study. The analysis itself involves reading the texts in a manner similar to that described in explanations of thematic analysis: The researcher engages in an iterative process of becoming familiar with and immersed in the data, searching for themes and meaningful patterns, and reviewing these patterns (Braun & Clarke, 2006). Of course, common topics themselves can easily become mere laundry lists (see Walsh, 2010). Therefore, just as the coded categories from thematic analysis need to be written into narrative, commonplaces need to be organized in a meaningful way. As McKeon (1987) notes, "data, or single terms, can be pointed to but meaningful discourse arises only when two or more terms are combined in interpretation and connections or arguments are constructed" (p. 6). A schematic, in other words, creates a picture of the relationships between these topics. McKeon's (1966) own schematic, as described in "Philosophic Semantics and Philosophic Inquiry," was developed for the discovery of meaning in all discourses, but a similar exercise may lead beyond compromise between individual motives to collective solutions for interdisciplinary problems in science communication.

Commonplaces in the 23andMe Controversy

The FDA, 23andMe, and 23andMe's customers share six commonplaces: (1) the need for understanding complex information, (2) the value of empowerment, (3) the importance of innovation, (4) the need for safety, (5) the importance of accurate, reliable information, and (6) the importance of privacy or confidentiality.

GETTING AND UNDERSTANDING COMPLEX INFORMATION

Both the FDA and 23andMe value translation or mediation as the most ethical way to communicate complex information to a lay public. 23andMe sends an evolving message about this need as its relationship with the FDA shifts. In 2007, the home page of the company's website was a job ad for science writers who could "create well written educational content for the general consumer" and could "convey complex scientific concepts in easily understandable prose for non-scientists." These science writers would act in conjunction with subject matter experts to "translate" complex data to a lay audience.

Later that same year the company seemed to change its stance on how information needed to be communicated to its customers, as evidenced by new home page text:

> Genetics is About to Get Personal. Don't panic, we're here to help. 23andMe is a privately held company developing new ways to help you make sense of your own genetic information. Even though your body contains trillions of copies of your genome, you've never likely never read any of it. Our goal is to connect you to the 23 paired volumes of your own genetic blueprint.

Relying on metaphors of "texts" and "blueprints," 23andMe begins to align its audience with the role of a novice being taught to read this new language. The implication seems to be that eventually a person could learn to do this on their own, much like people learning to read at a young age go on to do. In essence, 23andMe expresses values of mediating information to the public through metaphors and simplified language or through advising people to mediate the information by talking with their doctor. In May 2011, for example, the home page promoted a testimonial from a customer, Kirk C. According to Kirk's text, because he took his results from the 23andMe test to his doctor, his doctor "got to a diagnosis much faster. 23andMe saved my life." The message thread continued in the following year on the website:

- Take a more active role in managing your health
- Knowing how your genes may impact your health can help you plan for the future and personalize your healthcare with your doctor

A value of communication through mediation with a medical professional is also evident with other DTC testing companies. In fact, DNA Direct and Navigenics have begun to shift to a "direct-to-provider" (DTP) marketing strategy (McGowan, Fishman, Settersten, Lambrix, & Juengst, 2014). Such a strategy targets clinicians as consumers of products and product advertising and in doing so helps them retain roles as gatekeepers to personal genetic information even if the clinician may not specifically have expertise in the area of genetics.

These shifts in terms of the way technical information is mediated, from something done by an employee of the company (a science writer) to something done by customers' doctors, may be an artifact related to the conversations between such companies and the FDA, a conclusion supported by the fact that just six months before the FDA's warning letter the 23andMe home page said, "Learn valuable health & ancestry information." And of course the following year the company focused solely on the ancestry component. This change in emphasis with respect to what customers might learn from their genetic test results is consistent with the FDA's 2013 warning letter, which emphasizes information being medically mediated for better understanding. The letter states, "Serious concerns are raised if test results are not adequately understood by patients or if incorrect test results are reported" (n. p.).

Unlike the discourse on 23andMe's website, however, the FDA takes the stance that continual mediation with a qualified scientific or medical expert is required, even as the agency reversed its stance on 23andMe's product when it granted approval for the company's carrier status test for Bloom syndrome in February 2015. As of this date, parents were able to take this test to see whether they carry mutations that could cause a rare disorder called Bloom syndrome in their children. The agency said that, in general, such so-called carrier tests would no longer need to be approved in advance before being marketed. It also voiced support for allowing consumers in some situations to have direct access to genetic testing, without a doctor being involved. In a press release about this particular decision, the FDA states: "The FDA believes that in many circumstances it is not necessary for consumers to go through a licensed practitioner to have direct access to their personal genetic information" (2015, n. p.). At the same time, a value of having genetic test results communicated in a mediated manner is included in the press release as well. If sold over the counter (which can be interpreted as referring to items sold over the internet as well), the FDA requires 23andMe to provide information to consumers about how to "obtain access to a board-certified clinical molecular geneticist or equivalent to assist in pre- and post-test counseling" (FDA, 2015, n. p.). The implication of this directive is that if this medical condition exists and future decisions need to be made based on the test results, a medical professional should mediate the interpretation of the results.

The 73 reviews from customers examined for this project, on the other hand, indicate they do not place as high a value on such translation or interpretation, even though there is a sharp divide between comments that express satisfaction with the level of detail in the results 23andMe returned. Of the 33 reviewers that mention this topic, 13 were satisfied, stating, for example: "Again, as I say, I am very satisfied with my lab results. They were comprehensive and interesting—both the ancestral information and the health indicators" or "The amount of information provided by the test is unbelievable." The other 20 were not happy with the level of detail 23andMe sent back to them, making comments such as the following:

> So you send in your $99 and about three or four months later you get a very generalized and vague reporting of what continent(s) your ancestors came from. Lo and behold, most of us humans originally came from Africa if you go far enough back—something that just about any science book could tell you for free. Oh, and if you're a white male—you probably have some link to various parts of Europe prior to that. Wow. I also found both the so-called health analysis to be pretty much useless [sic]. Sure, you're told that people with your genetic markers have a higher percentage than others markers of a certain disease or health risk, but so what? That is totally meaningless as lifestyle, diet, exercise and a host of other factors come into play and serve a far more important determination of what kind of health you have and are going to have. I didn't expect a lot from this test—but perhaps something a little more defined than what this offers.

These customers indicate a desire for more specialized language rather than specialized information that is simplified or excluded from the reports they received. This particular customer, in fact, clearly resents the company's assessment of what he would find to be useful and as such suggests that the company underestimates its customers' ability to understand and interpret personal genetic information.

Access to information can lead to empowerment because people can make decisions and act on this information. As the previous excerpts from customers suggest, the 73 Amazon reviews were split between the information received being "just so vague" and specific, actionable information. In addition to expectations about the level of detail offered in the results, if the customer did not think the testing was valid or accurate, the person did not feel empowered by the information: "Well, to sum it up: if you do this test, you will be informed that you are related to EVERYBODY and if you are hoping to see just a very general summation: you will be ok, but anything other than knowing you are Asian, African, European will be no help. Save your money and go get a massage." Customers who did find the information accurate and useful, on the other hand, did feel empowered through the knowledge.

> On the health side, it gives you info about: your genetic predisposition and carrier status (or lack of) for diseases, how well your body will likely respond to certain drug treatments, and your probable genetic traits. On the ancestry side, it gives you info about: your ancestor's geographic regions, (if you opt-in and your family joins), you can compare DNA with relatives to see which lines you got which traits from, (if you opt-in) you can also find (and, if you like, contact) close or distant relatives. It's awesome for those trying to find out more about their genealogy. I found the ancestry stuff totally fascinating and the health stuff very useful. WORTH EVERY PENNY.

The company 23andMe has always heralded customer empowerment on its website and in discussions with the FDA. In her presentation to the FDA in the public meeting that took place in March 2011, Ashley Gould, 23andMe's General Counsel, stated: "We firmly believe that individuals have a fundamental right to directly access information about their own DNA. Empowering people to become informed healthcare consumers is critically important to making the widespread practice of personalized medicine a reality." One way the commonplace of empowerment is expressed by 23andMe is that innovation is seen as a process of science and science policy that is participatory.

The company has also seen participation as a key principle to genetic testing scientific research. Looking at non–health-related DNA testing, Nordgren (2010) describes this commonplace as related to items that fall on a spectrum between individualistic and communitarian concerns. Individualistic concerns stress that each individual is unique. The communitarian vision emphasizes that individuals are members of communities, in this case genetic communities. This community ethic is clear in the company's 2009 research project with Parkinson disease. In the section of the company's website that explains the research project the text encourages the reader to "take an active role in groundbreaking research by mailing in your DNA sample and answering surveys online." But by 2010 this theme shifted from a more paternalist stance of the company helping the customer to the customer being in control by taking action. The empowerment theme is evident in text such as "choose the DNA test that's right for you," "take charge of your health," and "choose to have it all." The need for action is also apparent in the directives "choose" and "take charge." This "doing" ethos continued through 2011 as the site was actually transformed into a set of instructions with steps. These steps are: "1 Get Your Kit," "2 Provide Your Saliva," "3 Learn About Yourself," and "4 Get Monthly DNA Discoveries." The website incarnation in 2015, however, stresses individual empowerment through the individual's choice to participate or not. It includes everything from the tagline "A new way to see yourself. Through your DNA," to the page that includes individuals' stories, to the Research page that says:

> You can make a difference by participating in a new kind of research—online, from anywhere.
> — Understand what your DNA says about your health, traits and ancestry
> — Share and compare with tools to engage family and friends
> — Receive ongoing reports as new genetic discoveries are made and as we are able to clear new reports through the FDA
> As a customer, you can answer online survey questions, which researchers can link to your genetic data to study topics from ancestry to traits to disease. Your contribution helps drive scientific discoveries. You can always choose to opt into or out of research.

Here, empowerment is expressed as driving important research, and the value of having the freedom to do what one wants to with their own data is clear. At the same time, the consumerist discourse is strongly threaded through the site with an "Order Now" button on every page except the Stories page.

As a term, the commonplace of participation is difficult to uncover in the FDA texts, but their process itself is participatory: The FDA invited DTC genetic testing companies to come and talk with them, holding public hearings that included not only representatives from the FDA and 23andMe but also customers and medical professionals. Although it is less apparent in the

FDA's initial communications with 23andMe, the agency seems to hold a similar view in its 2015 communication about the approval of the Bloom syndrome test, which orders: "explain to the consumer in the product labeling what the results might mean for prospective parents interested in seeing if they carry a genetic disorder" (FDA, 2015, n.p.). In this statement and in their action to let 23andMe offer this testing, the FDA is signaling that they value access to this information.

Additional commonplaces seek to balance safety and technological innovation. For both the FDA and 23andMe, consumer safety is a large part of reliability, but it must be balanced with innovation. Elizabeth Mansfield, Officer of In Vitro Diagnostic Devices at the FDA, expressed this concern in the 2011 Advisory Committee meeting: "In addition, the regulatory apparatus must keep pace with rapidly advancing technology and scientific knowledge, as discussed on the previous slide. We must be able to assess new technologies and promote high-quality innovation, while protecting patients" (2011, n.p.). The position of 23andMe on innovation can be found both on the company's website and in the transcription of the presentation by Ashley Gould at the Advisory Committee meeting, in which she stated:

> In conclusion, I leave you with 23andMe's requests for your consideration: First, continue to allow informed consumers to freely learn about their own DNA. Adopt *thoughtful policy* that promotes innovation and is flexible enough to evolve with new technologies and research developments. . . . Through a cross-sector working group, effectively define *clinical validity* specific to genetic testing. Finally, focus on *establishing requirements* for analytical and clinical validity, analytical standards and transparency that apply to all genetic testing services. Genetic information provided directly to consumers should be held to the same standards as genetic information provided in a clinical setting. (emphasis in original)

At the 2011 Advisory Committee meeting Dan Vorhaus, an attorney with Robinson, Bradshaw & Hinson, expressed a similar need for this balance:

> And the real tensions here are, I think, twofold. One is innovation tensions of balancing public health concerns and concerns for consumer safety against innovation and concerns for consumer access and desire for direct access. And then, again, this fundamental tension of interposing clinical guidance into the process of DTC genetic testing as opposed to safeguarding individual autonomy and giving people the ability to access that information on their own.

As with the commonplaces associated with access to information, the commonplaces of safety and technological innovation are edged with values of mediation. Whereas in terms of communication mediation is seen as enhancing understanding, here mediation is expressed in terms of guiding actions. This can be illustrated through comments Shuren made in his testimony to Congress: "Marketing genetic tests directly to consumers can increase the risk of a test because a patient may make a decision that adversely affects their health, such as stopping or changing the dose of a medication or continuing an unhealthy lifestyle, without the intervention of a learned intermediary." Mediated communication, in other words, can benefit not only in terms of understanding but also in safe, ethical action whether this mediation is from a government agency, the company, a medical professional, or early adopters of the technology.

Many of the customers who bought 23andMe's testing kit on Amazon and posted reviews of the product also embrace technological innovation but in doing so seem to place less stress on mediated actions: "The 23andMe Genetic test is well worth the money. The product will become even more useful as more people are tested. It is the beginning of a new era in medical technology." Some of these users represent their role in this process as that of an early adopter:

> I don't expect a doctor to know everything or tell me everything about my body, because that isn't possible. Why expect perfection from an early effort at genetics? We are early adopters who realize the potential. As early adopters we were a group mostly of educated professionals. We were not rubes who had to be protected. For me this test was great fun! I didn't expect to be wowed! But I was and still am.

Potential adopters of a technology look to early adopters for advice and information (Rogers, 2003). As such, early adopters of technologies are often characterized as opinion leaders and become mediators of information themselves. They try a technological innovation and then communicate the usefulness of the new product to other consumers. In the Amazon reviews, early adopters' function as opinion leaders can be identified through the feedback from other customers in the form of how many people found a review helpful. Nine of 11 people who read the previous review found it helpful. In other words, early adopters advise people. One particular way they advised other customers in the 73 reviews examined here was by directing people to third parties to interpret the raw genetic data they received from 23andMe. According to the Amazon customer reviews, consumers see this third-party source as an objective and machine-like interpretation that is valid and accurate. The experiences with third parties were described as positive experiences.

Some customers of 23andMe who purchased the testing kit through Amazon also self-identify as experts. One indication that can be pointed to for such self-identification is the use of the link "See all my reviews." This link might suggest that rather than simply ranking a product on occasion, the reviewer sees reviewing as a responsibility she has for the community. In a sense, she might consider herself a "professional" reviewer of sorts. Amazon encourages this type of self-identification through its Top Reviewer Ranking list and its Hall of Fame Reviewers list. The Top Reviewer list consists of the "best" contributors at the present time. The Hall of Fame list includes reviewers who had been highly ranked in previous years. Rankings are based on statistics for total reviews and the number of people who claimed to find the review helpful. In the reviews examined for the 23andMe test kit, expertise is also claimed by some customers through a preference for receiving non-mediated test results. Three of the reviewers mentioned preferring the raw data to an interpretation of the raw data, as the following excerpts illustrate:

> Excerpt 1: The real value of 23andMe is the raw data. If you have something that runs in your family, knowing your genes and comparing it to real medical studies, is the best way to go. The studies on 23andMe are limited. The other feature that I found really interesting is the chances for side effects of popular medications. It was spot-on with a side effect I had in the past with a medication I took for a while.

> Excerpt 2: Having my raw data has allowed me to look further into a genetic condition that runs in my family but isn't something that 23andMe specifically tests for yet. I had a 50/50 chance of inheriting something I didn't want. Now I know I lost that particular genetic lottery. Environmental and lifestyle factors may still keep the disease from manifesting but, armed with the knowledge that it is a real possibility, I can keep a close eye out for symptoms and not make the mistake of writing them off to normal aging. Some people would not want to know, but I did. I can now undergo regular testing, which will allow me to start treatment at the earliest detectable stage of the disease, should it manifest. My doctor tells me that the standard treatment is to first take a "wait and see" approach but, in my case, he won't be doing that. He's delighted to have the knowledge he needs to treat me more effectively, should the need arise.

As Condit (2010) points out, many people do believe that health conditions are influenced by both genes and other factors like the environment. Along with that understanding, the customer in the second excerpt not only gets and works with non-mediated, raw data, he also exhibits expertise by act-ing much like a medical professional might, using the intervention of watch-ful waiting—an intervention that gives a problem time to resolve before per-forming additional tests or undertaking more invasive interventions.

RELIABILITY AND ACCURACY

Actionable information also requires the commonplace of reliability, which is linked to the accuracy of the test results and the company's credibility. 23andMe consistently defines the company's reliability through its affiliations with scientists and the company's professed adherence to scientific standards and methods. The two areas of the website where this is most apparent is the page entitled "Scientific Standards" and the publication of the company's advisory board and the board members' biographies. On the Scientific Stan-dards page, for example, 23andMe lists six standards: 1) laboratory process-ing, 2) development and curation of scientific databases, 3) a rigorous review process, 4) scientific and software innovation, 5) expert advice, and 6) sci-entific progress. Interestingly, the introduction to these standards seems to weave business concerns of reliability such as "quality control" with scientific concerns such as validity and accuracy. The introduction states: "Our scien-tific processes have been designed with the goal of providing our customers with accurate, high-quality data in a format that is easy to understand and interpret. Each step in our workflow is carefully monitored and validated through quality control measures that ensure your data accurately reflects your genetic makeup."

This accuracy is linked to analytical validity, which is part of the frame-work for genetic test evaluation. The framework takes its name, ACCE, from the four components evaluated: analytical validity; clinical validity; clinical utility; and the ethical, legal, and social implications of genetic test-ing (Kroese, Elles, & Zimmern, 2007). According to that discussion, the analytical validity of a genetic test defines its ability to measure accurately and reliably the genotype of interest. In response to this need, in 2008 the 23andMe website also contained this statement: "The genotyping services of 23andMe are performed in LabCorp's CLIA-registered laboratory. The results presented in Health and Traits have not been cleared or approved by the FDA but have been analytically validated according to CLIA standards." In his tes-timony to Congress in 2010, Jeffrey Shuren similarly discussed the issue of reliability in terms of analytical validity: "Premarket review of moderate and high risk LDTs [laboratory developed tests] would ensure that the tests are evaluated for analytical validity and clinical validity, based on their claimed intended use, and would provide an independent and unbiased assessment of the data used to support analytical and clinical claims for those LDTs" (n. p.).

The customers on Amazon reviewed here, however, define reliability in terms of test results that match their expectations:

> Next, the results were outstanding. The paternal and maternal haplogroups matched exactly with familytreedna results including deep clade classification. The test results also matched by blood type correctly and accurately predicted traits like eye colour, lactose intolerance, alcohol flush reaction and others. So the test results are sound and the information provided with the results is excellent.

In other words, unlike the company and the FDA, for customers the reliability of the test results is at least partially assessed through their own personal knowledge of their health status or family history.

Reliable test results were also linked to customer service issues for customers of 23andMe. All the customer reviews indicated three variables that influence perceptions of 23andMe itself as being a reliable company: (1) price fairness, (2) transparency (in terms of the company's use of customer data), and (3) customer service. Customer discourse focusing on validity most often discussed the quality of the test results: "Afterwards I conducted many researches by geneticists on the internet that broadly concludes that these 'take at home kits' are not as reliable as people would like them to be. I wash my hands from these tests now and will continue the old fashion way, family telling me and genealogical records." Customers also related validity to the company's credibility and reliability as a business: "This seems to be the best DNA service out there. It has at least as many features as other consumer DNA services and presents the info in a VERY user-friendly way. Plus, they add new features all the time."

Many of the customer reviews on Amazon voiced frustration at lengthy waits for results. Often, these waits were also linked to expectations about the amount of information the results would contain. As one reviewer put it: "Too long to receive too general information. After waiting over 6 weeks, I learned that I'm half Southern European, and half Northern European. Really? I knew that but what I paid for was what specific origins I had. What a waste of money!!!" Another customer stated:

> This company is obviously investing the money it gets in interest bearing accounts so it can make extra profit on what consumers spend for the extra 5–6 weeks they take to send you your results. How long does it take to run a sample in a machine? It does not take 2 months. My friend works at Genetic.

She said they are obviously delaying the process on purpose as 1–2 weeks is tops here. I can't stand companies that legally rip off consumers on purpose. Here is another one.

Customer comments like this one articulate the themes of what the customer sees as a simple process for completing the genetic test and expectations of associated quick shipping times. Nine customer reviews discussed issues of time; these also described the inability to reach a customer service representative in a timely manner as a reason they had negative perceptions of the company's credibility.

PRIVACY AND CONFIDENTIALITY

For customers, unlike the FDA and 23andMe, privacy and confidentiality of information is more important than participating in the scientific discovery process (not the process of a person's individual self-discovery). This comes across most clearly in the customer reviews that link trust about the accuracy of the data to the amount of time it took to get their results and to the company's customer services (which was rated poorly by all the reviewers). One reviewer, for example, said:

> Basically, I can't say enough bad things about this company and hate to even give them 1 star. And I went into this so supportive of 23andme and their mission. And this is a company that holds what is perhaps some of your most personal biological information (along with health history, browsing data, whatever else their cookies track . . . they are backed by Google). If they treat their customers/research participants this poorly, than I can't imagine they have a culture of respecting customer/participant data and making reasonable, legal and ethical decisions when it comes it sharing it. Also, if they cannot get their basic order processing/ analysis flow under control, I question their ability to handle sensitive data and give accurate analysis. They blame their partner lab NGI, but they seem excellent. 23andme is the problem.

Clearly, 23andMe's commitment to research and to participatory, open-source research needs to be balanced by such concerns some purchasers of the company's DTC test kit have about credibility, privacy, and confidentiality.

Building Ethical Frameworks from Commonplaces
to Find an Ethical Way Forward

The plurality that evolves from uncovering commonplaces in DTC genetic testing communication can be messy. Definitions overlap and boundaries blur, but taking the next step in analysis, building a schematic of these commonplaces, can help us see these relationships more clearly, and understanding these relationships is useful in communicating to diverse stakeholders. In this schematic, terms for commonplaces are combined and aligned with interpretations of the topics from the point of view of each stakeholder. As a whole, the terms, their connections, and interpretations anchor an ethical framework.

In examining the texts from 23andMe, the FDA, and 23andMe customers, I attempt an early version of a schematic (table 11.1) based on these ideas about applied ethics and McKeon's work with commonplaces. To create this schematic I use the commonplaces I identified as a starting point and note the value each stakeholder attaches to the commonplace. Finally, I assign one of the three ethical standpoints McKeon discusses: 1) an ethics of virtues related to considerations of the nature, powers, and ends of men, 2) an ethics of precepts related to the duties and obligations of man relative to himself and to others, or 3) an ethics of responsibility (McKeon, 2005). In his discussion of ethics, R. McKeon (2005) highlights an ethics of virtues that is concerned with what "is" or facts; an ethics of precepts, which focuses on what individuals should do for the public good; and an ethics of responsibility, a framework that "is determined by reciprocities in the actions of men, and which relates virtues and precepts preserving their differences in their indifference" (p. 252). We have "rules" of nature and rules of law. Ethical frameworks and language in these stances exist in science and government regulation/oversight. The third framework, the one of responsibility, as I interpret McKeon, refers to the individual person's actions within a society that has rules of both nature and law.

As the schematic in table 11.1 suggests, there are tensions and agreements in what the stakeholders value and what ethical framework informs their communicative actions in this scientific controversy. The commonplace of expertise or the need to understand complex information, for example, highlights disparities between all three stakeholders as well as the overlaps. The FDA and 23andMe value mediated information. The company 23andMe and the customers in this study also value the customer's role in being the person mediating this information or not needing anyone to interpret or translate the information they receive in genetic test results. The commonplace of innovation is valued by all of the stakeholders, but the FDA views innovation as a "threat" to safety and privacy. Some customers see themselves as innovators in their role as early adopters, and 23andMe views innovation as participating in research or having the choice to participate in research.

TABLE 11.1. An Ethical Schematic for Communicating DTC Genetic Testing Results

Commonplace	Value Expressed	Ethical Principle	Stakeholder
Need to understand complex information	Customer needs translator	Precepts	FDA and 23andMe
	Customer is translator for self	Responsibility	23andMe and customer
	Customer is translator for other customers	Virtue	Customer
Value of empowerment	Testing allows customer to participate in scientific research	Responsibility	23and Me, Customer, and FDA
	Education (learning the results) is empowering in terms of decision making	Virtue	Customer
Importance of innovation	It is important to be an early adopter of technology to help explain to other adopters	Responsibility	Customer
	It is important to participate in scientific discovery as a citizen	Responsibility	23andMe
	Acts of innovation should be balanced with concerns about safety	Responsibility	FDA
Need for safety	Public/government agencies have the responsibility for securing safety	Virtue	FDA
	Action based on information may be risky to individuals' health	Responsibility	FDA
	Actions based on test results that are balanced with personal knowledge can be acts of health prevention	Responsibility	Customer
Need for accurate, reliable information	Reliability equals analytical validity	Precepts	FDA and 23andMe
	Reliability equals honesty	Virtue	Customer
	Reliability equals the technology working right	Precepts	Customer and FDA
Importance of privacy or confidentiality	Privacy of the owner of data should be respected	Virtue	Customer
	Business and scientific processes of the company (23andMe) should be transparent	Responsibility	Customer and FDA

Conclusion

Communication in the landscape of DTC genetic testing is an interdisciplinary endeavor in that it involves science, government, and technical and public spheres. And yet, efforts to communicate effectively, accurately, and ethically have largely been framed through a "top down" or even a "bottom up" communication model such as discussed by Felt et al. (2009). Discussions of ethical DTC genetic testing communication, however, need to move beyond static conversations about which direction information flows from and apply a flexible communication framework based on understanding and applying the multiple interests of interested parties.

One way to do this is by mapping out the common topics and the associated values and ethical frameworks in such a complex situation. By sorting out the common topics for each stakeholder and identifying the ethical framework driving stakeholders' interpretations of these topics, we come to a place for proactive solutions rather than reactionary actions for communicating in such situations. As Jeremy Gruber, president of the Council for Responsible Genetics, said at the 2011 Advisory Committee meeting:

> As the science progresses, it has become clear that the major challenge for the future won't be sequencing technologies and broad public access to them, but rather the cost and difficulty of interpreting and applying the huge amounts of data that they generate. We are still only at the beginning of this genetic revolution, and it's certainly our hope that this new synthesis of genetics and information technology can empower individual self-knowledge and promote health access across a wide variety of platforms.

Issues of communication and interpretation will be at the center of these challenges. As such, science communicators will continue to play a pivotal role in the complex field of genetic testing. This role can be one that helps generate the questions that inform actions and policy.

References

Braun, V., & Clarke, V. (2006). Using thematic analysis in psychology. *Qualitative Research in Psychology, 3*(2), 77–101.

Condit, C. M. (1999). *The meanings of the gene: Public debates about human heredity.* Madison: University of Wisconsin Press.

Condit, C. M. (2010). Public understandings of genetics and health. *Clinical Genetics, 77*(1), 1–9.

Critchley, C., Nicol, D., Otlowski, M., & Chalmers, D. (2015). Public reaction to direct-to-consumer online genetic tests: Comparing attitudes, trust and intentions across commercial and conventional providers. *Public Understanding of Science, 24*(6), 731–750.

Curnutte, M., & Testa, G. (2012). Consuming genomes: Scientific and social innovation in direct-to-consumer genetic testing. *New Genetics and Society, 31*(2), 159–181.

Delfanti, A. (2011). Know your genes: The marketing of direct-to-consumer genetic testing. *Journal of Science Communication, 10*(3), 1–2.

Einsiedel, E. F., & Geransar, R. (2009). Framing genetic risk: Trust and credibility markers in online direct-to-consumer advertising for genetic testing. *New Genetics and Society, 28*(4), 339–362.

Felt, U., Fochler, M., Müller, A., & Strassnig, M. (2009). Unruly ethics: On the difficulties of a bottom-up approach to ethics in the field of genomics. *Public Understanding of Science, 18*(3), 354–371.

Global Industry Analysts. (2012, June 10). Future of direct-to-consumer (DTC) genetic testing market remains fraught. Retrieved from http://www.prweb.com/releases/DTC_genetic _testing/direct_to_consumer_tests/prweb9780295.htm

Gronnvoll, M., & Landau, J. (2010). From viruses to Russian roulette to dance: A rhetorical critique and creation of genetic metaphors. *Rhetoric Society Quarterly, 40*(1), 46–70.

Gould, A. (2011, March 8). Comment made during public meeting of the Medical Devices Advisory Committee on the Molecular and Clinical Genetics Panel, Washington, DC.

Harris, A., Wyatt, S., & Kelly, S. E. (2013). The gift of spit (and the obligation to return it). *Information, Communication & Society, 16*(2), 236–257.

Javitt, G. H., Stanley, E., & Hudson, K. (2004). Direct-to-consumer genetic tests, government oversight, and the first amendment: What the government can (and can't) do to protect the public's health. *Oklahoma Law Review, 57*, 251–302.

Kroese, M., Elles, R., & Zimmern, R. (2007). *The evaluation of clinical validity and clinical utility of genetic tests.* Cambridge, UK: PHG Foundation.

Lippman, A. (1992). Led (astray) by genetic maps: The cartography of the human genome and health care. *Social Science & Medicine, 35*(12), 1469–1476.

Lopez, J. J. (2007). Notes on metaphors, notes as metaphors: The genome as musical spectacle. *Science Communication, 29*(1), 7–34.

Mansfield, E. (2011, March 8). Comment made during public meeting of the Medical Devices Advisory Committee on the Molecular and Clinical Genetics Panel, Washington, DC.

McGowan , M. L., Fishman, J. R., Settersten Jr., R. A., Lambrix, M. A., & Juengst, E. T. (2014). Gatekeepers or intermediaries? The role of clinicians in commercial genomic testing. *PLOS ONE, 9*(9), e108484.

McKeon, R. (1966). Philosophic semantics and philosophic inquiry. Retrieved from http://www .richardmckeon.org/content/e-Publications/e-OnPhilosophy/McK-PhilosophicSemantics &Inquiry.pdf

McKeon, R. (1987). *Rhetoric: Essays in invention and discovery.* Woodbridge, CT: Ox Bow Press.

McKeon, R. (2005). *Selected writings of Richard McKeon: Vol. 2. Culture, education and the arts.* Chicago, IL: University of Chicago Press.

Miller, C. R., & Selzer, J. (1985). Special topics of argument in engineering reports. In L. Odell and D. Goswami (Eds.), *Writing in nonacademic settings* (pp. 309–341). New York, NY: Guilford.

Murphy, P., & Maynard, M. (2000). Framing the genetic testing issue: Discourse and cultural clashes among policy communities. *Science Communication, 22*(2), 133–153.

Nelkin, D. (2001). Molecular metaphors: The gene in popular discourse. *Nature Reviews Genetics, 2*, 555–559.

Nordgren, A. (2010). The rhetoric appeal to identity on websites of companies offering non-health-related DNA testing. *Identity in the Information Society, 3*(3), 473–487.

Pearson, Y. E., & Liu-Thompkins, Y. (2012). Consuming direct-to-consumer genetic tests: The role of genetic literacy and knowledge calibration. *Journal of Public Policy & Marketing, 31*(1), 42–57.

Perez, G. K., Cruess, D. G., Cruess, S., Brewer, M., Stroop, J., Schwartz, R., & Greenstein, R. (2011). Attitudes toward direct-to-consumer advertisements and online genetic testing among high-risk women participating in a hereditary cancer clinic. *Journal of Health Communication, 16*(6), 607–628.

Rogers, E. M. (2003). *Diffusion of innovations* (5th ed.). New York, NY: Free Press.

Ross, D. G. (2013). Common topics and commonplaces of environmental rhetoric. *Written Communication, 30*(1), 91–131.

Rothman, B. K. (1998). *Genetic maps and human imaginations.* New York, NY: W. W. Norton.

Seife, C. (2013, November 27). 23andMe is terrifying, but not for the reasons the FDA thinks. *Scientific American.* Retrieved from http://www.scientificamerican.com/

Shuren, J. (2010, July 22). Comment made during Congressional testimony on Direct-to-Consumer Genetic Testing and the Consequences to the Public, Washington, DC.

Spindle, D. (2013, November 25). 23andMe: Controversial genetic test kit tells background, future risks. ABC15. Retrieved from http://www.abc15.com

US Food and Drug Administration (FDA), Inspections, Compliance, Enforcement, and Criminal Investigations. (2013, November 22). 23andME warning letter (Document Number GEN1300666). Retrieved from http://www.fda.gov

US Food and Drug Administration (FDA), Inspections, Compliance, Enforcement, and Criminal Investigations. (2014). Is the product a medical device? Retrieved October 3, 2015, from http://www.fda.gov/

U.S. Food and Drug Administration (FDA). (2015, February 19). FDA permits marketing of first direct-to-consumer genetic carrier test for Bloom syndrome. Retrieved from http://www.fda.gov

Vorhaus, D. (2010, August 5). The past, present and future of DTC genetic testing regulation. Genomics Law Report. Retrieved from http://www.genomicslawreport.com

Vorhaus, D. (2011, March 8). Comment made during public meeting of the Medical Devices Advisory Committee on the Molecular and Clinical Genetics Panel, Washington, DC.

Walsh, L. (2010). The common topoi of STEM discourse: An apologia and methodological proposal, with pilot survey. *Written Communication, 27*(1), 120–156.

William-Jones, B. (2006). Be ready against cancer, now: Direct-to-consumer advertising for genetic testing. *New Genetics & Society, 25*(1), 89–107.

Wilson, J. C. (2002). (Re)writing the genetic body-text: Disability, textuality, and the Human Genome Project. *Cultural Critique, 50,* 23–39.

12

Excluding "Anti-biotech" Activists from Canadian Agri-Food Policy Making: Ethical Implications of the Deficit Model of Science Communication

KELLY BRONSON

On a 2015 episode of the Canadian debate show *The Agenda*, the host, Steve Paikan, described what he called "an enduring conflict" between fact and belief. Paikan asked the panelists (of which I was one), "Why do nonscientists continue to believe certain things even when science has shown them to be false?" This chapter will make the broad argument that this dominant "deficit model" framing of science-public discord—publics as ignorant of the facts given by science—is quite often a faulty and unproductive framing that forestalls public engagements with science that could otherwise enrich its democratic importance. Most importantly for the purposes of this book, when science communicators—whether scientists, TV show hosts, or policy makers—characterize resistant publics as merely ignorant, they are not simply making an empirical mistake; they are also defining critical public engagements with science as problematic and justifying the exclusion of members of the public from active participation in political processes involving science and technology. Thus, the decision to use the deficit model has ethical implications.

This chapter uses original empirical research with farmers and activists in Saskatchewan, Canada, to highlight and explain how their exclusion from regulatory decision making around agricultural biotechnologies[1] in Canada was furthered by assumptions made among governance actors that ignorance was at root of technological resistance. I offer an alternative framing; by

---

1. The term *biotechnology*, or biotech, is used here to refer to the process of transferring genes (and regulatory elements) among organisms in the production of new ones and in a laboratory space, as well as the products of this lab work. Farmers sometimes referred to GM or genetic modification, which is the process resulting in biotechnologies.

drawing on qualitative data gathered over a three-year period (2001–2004), I reveal the complexity of concerns motivating this public. Saskatchewan farmer resistance to biotechnologies springs from a deep and local history of relationships that have come to characterize the Canadian food system. Resistance also springs in this case from a critique of the values embedded in commercial biotechnological science. The study hopefully provides an opening for science communicators of all kinds to think carefully about the ethical consequences of continued adherence to the dominant way of framing science-public tensions over biotechnologies—publics as "anti-scientific"— for the ongoing struggle over its knowledge-power structures.

Dominant Deficit Understandings of Biotechnology Resistance

A popular narrative on the subject of biotechnologies is that "the public" is deficient in technical know-how and thus susceptible to a general "anti-biotech" attitude (see Lynas, 2015). Public ignorance persists as a widespread explanation for why biotechnology resistance endures in Canada two decades after the initial commercialization of agricultural biotechnologies (notably seeds altered to work with specific chemical herbicides). Considerable efforts to promote biotechnologies by educating publics about the science behind them have failed to completely secure these technologies as unproblematic in Canada, or for that matter elsewhere (Britain, Argentina) where they have been deployed (Byrne et al., 2002; Knight, 2009). Case studies have revealed that Canadian biotechnology decision makers (namely, regulators) define those members of the public who are critical of biotechnologies as ignorant in technical matters, as anti-science, and therefore as peripheral to governance processes (Shields, 2008; Montpetit & Rouillard, 2008). Government workers have, for example, described critical publics as "passive, emotional and ignorant" (Cook, 2004, p. 38; see also Cook & Robbins, 2002; Priest, 2000; Gregory & Miller, 1998). On the basis of interviews with Canadian regulatory scientists, Montpetit and Rouillard (2008) suggest that within the culture of Canadian biotechnology risk assessment, "science is treated as monolithic and believed to produce universal truth and the perspectives of social groups on truth matters little in risk-management processes. Social groups are then treated, pejoratively, as special interest groups" (p. 915).

In interviews, regulatory actors appeal to science as something fixed, as secure knowledge that "the public" lacks and needs and that if "the public" was only more rational it would take up without concern (Bora, 2010; Irwin & Wynne, 1996). In the words of one Canadian biotechnology regulator, oppositional publics are justifiably ignored in policy processes because they hold "irrational fears" about accepted scientific facts and attending to such anti-science would be "pandering to rabid democracy" (Shields, 2008, p. 133). Lorne Hepworth—the spokesperson for CropLife and thus all biotechnology corporations in Canada—expounds the deficit view in this online response to a group of people concerned about biotechnology seeds and asking online for a moratorium:

> Health Canada has said for years that [genetically engineered technologies] can be safely used, and this is determined by the most modern scientific information. . . . [GE technologies] are some of the most stringently regulated products in Canada and only those products that meet Health Canada's strict health and safety standards are registered for sale and use. . . . The reality is that it simply isn't logical to ban these products.

A few assumptions appear to color Hepworth's response. First, Hepworth suggests that this group's concerns have to do with science per se, because it is assumed that they will be assuaged with the knowledge that biotech seeds are subject to governmental regulation informed by "modern" science. Second, Hepworth implies that maintaining a critical view on biotechnologies, given the scientific facts, is illogical. Rather than idiosyncratic, this quotation arguably reflects the dominant way of understanding resistance to biotechnologies in Canada.

This thinking is arguably framed by what is called "the deficit model" of science communication (for detailed deficit model histories see Hart & Nisbet, 2012; Brossard & Lewenstein, 2010; Gregory & Locke, 2008). Brian Wynne described the deficit model in 1991, suggesting that scientists and policy experts commonly assert that the reason why people disagree with them about the acceptability of technologies is because of public ignorance. Wynne took for granted that on much technical terrain typical publics know less than highly specialized experts, but at the same time he suggested that this is not typically why publics diverge (if they do) from these experts on the normative questions of social acceptability. The conception of publics as deficient in technical knowledge was codified by early (and earnest) attempts to respond ethically to intense public contestations over science-related health and environmental issues in the twentieth century. Protests over scientific and policy positions on issues such as environmental toxins and nuclear waste and weaponry were, by the late twentieth century, inviting speculation by concerned officials regarding both the source of and the solution to moments of social and political disruption. Given the nature of democratic systems, a lack of trust among citizens represents a legitimation crisis for policy makers intent on forming technical policy decisions that achieve a degree of social license, or a level of social acceptance (Prno & Slocombe, 2012).

In 1985, the Royal Society of London published a now-landmark report titled "The Public Understanding of Science" (known as the Bodmer report), which captured a more widely circulating belief that the public's interest in and support for science could be remedied if more scientists were to communicate their knowledge to the wider public. Bodmer was motivated by the laudable aim to make science more transparent and responsive to its publics. After the Bodmer report[2] an organized effort to promote public knowledge of science, and thereby instill confidence in it, was begun under the rubric of the deficit model (Turner, 2008). In 1992, the peer-reviewed journal *Public Understanding of Science* was begun to capture scholarly efforts to address the social ills surrounding public, and increasingly corporate, science. Since the time of Bodmer in the United Kingdom, the Canadian government has assembled various portfolios on science and society intended to increase public understanding of science; for example, the Canadian Biotechnology Secretariat began in 2006 with the intention of addressing the "crisis," as they describe it, of lack of public confidence in biotechnologies (a crisis, given the heavy public investment in such technologies since 1983).

In 2001, I set out to investigate the legitimacy of this deficit model framing of public resistance to agricultural biotechnologies, specifically that occurring on the Canadian Prairies—a hotbed of biotechnology controversy at that time. Unlike quantitative studies that test a generic public's scientific literacy in an attempt to explain it (e.g., Durant, Evans, & Thomas, 1989), I attempted to free myself from deficit model assumptions in order to hopefully shed new light on biotechnology protest (for similar case study work see Davies, 2009; Nisbet & Scheufele, 2009; Irwin & Michael, 2003; Priest, 2000; Wynne, 1991). My case study is framed more broadly by a theoretical commitment to seeing what counts as science or fact as contestable, open and shaped by the social context within which it is pursued (Irwin & Michael, 2003, p. 20; see also Bijker et al., 2009; Yearly, 2005; Sismondo, 2004; Haraway, 1995). Facts take shape and they achieve social stability—acceptance as truthful—in part through acts of communication qua legitimation processes, including through public debates. In this sense, attending to the mechanisms of science communication circulating around public debates are of incredible importance as they help to produce legitimation.

2. The US National Commission on Excellence in Education published a similar report in 1983.

Method

The claims of this chapter are supported by data drawn from an ethnographic study engaging farmers who were organized into legal action against biotechnology corporations in [1998] *Schmeiser v. Monsanto* and [2002] *Hoffman et al. v. Monsanto*. The participants were all characterized by their avoidance of crop biotechnologies (which placed them in a minority group in 2002) and most of them were organic farmers, though in 2002 there did not yet exist an enforceable national organic certification standard in Canada (Government of Canada, 2013).[3] My research goal in the original ethnographic study was to explore and shed light on the nature of this public's resistance to biotechnologies as well as to detail the culture of farming in an era of high technologies (Bronson, 2009). I built the data over many hours spent in farmers' fields, at farmhouses, at protests, and at a number of local "coffee row" gathering places, engaging farmers in unstructured and in-depth interviews that sometimes lasted days and which were guided by open-ended questions (e.g., "What are your thoughts about biotechnologies?").[4]

Seeing beyond Deficit Visions of Biotechnology Resistance

Saskatchewan farm fields, like farm fields around the world, are significant sites for the resistance of agricultural biotechnologies. Jose Bové was one of the first farmers to protest against biotechnologies by destroying biotech rice at the Nérac research field test sites of France in 1999, which action he explained as a way to bring the cultural consequences of biotechnologies into public consciousness (Hargrove, 2001). Peasant activists in less developed countries have prevented the planting of biotech seed systems through direct action protest, and they refuse to deploy biotechnologies as acts of protest against what they see as corporate profiteering at the expense of peasant livelihoods (SwissInfo, 2007). On the basis of my interviews, Saskatchewan grain

---

3. A national standard entitled Organic Agriculture (CAN/CGSB-32.310) was published by the Canadian General Standards Board (CGSB) in June 1999, but it was not an enforceable standard; enforceability did not come about until 2006: "Two national standards were published in September 2006: CAN/CGSB-32.310–2006, Organic Production Systems—General Principles and Management Standards, and CAN/CGSB-32.311–2006, Organic Production Systems—Permitted Substances List." Perhaps more influential, since all standards are voluntary, was the Canadian Food Inspection Agency's 2006 Organic Products Regulations.

4. For detail on the specific methodology (for sampling and transcription, for example) see Bronson, 2014a.

farmers fit into these farmer and peasant actions of resistance to agricultural biotechnologies for reasons that are social, political, and cultural.

CANADIAN BIOTECHNOLOGY RESISTANCE
IN ITS HISTORICAL CONTEXT

This study's data suggest that the social resistance to biotechnologies occurring on the Canadian Prairie must be understood within larger and historic social conflicts, such as that over different visions of good farming—intensive, aimed at export markets, versus farming that is geared toward environmental and community longevity. In this and the sections below, I attempt to make visible farmer perspectives in their contextual detail and specificity. All quotations are from farmers unless otherwise specified by in-text citation. Farmers' names are changed except for the case of two farmers who asked to be attributed.

Most of the farmers I interviewed had moved away from chemical applications in the mid-twentieth century out of environmental and social concern, but also because producing organic food for local markets builds sustainable social systems. "We were all really excited about new [ecological] ideas," "Jim" remembered about the emergence of environmentalism and ecological farming practice in 1970s Saskatchewan. Farmers described an upswing in popularity for organic production as it had occurred in the 1960s and 1970s in Saskatchewan, a time when there was new enthusiasm for experimentation in all kinds of alternatives and there was a burgeoning environmental politics. In his history of ecological thinking, *Nature's Economy*, Donald Worster has described the 1970s as North America's Age of Ecology. Similarly, other studies have shown that a number of farmers in North America adopted low-input agriculture during the 1970s owing to a rise in ecological insight—a rise in seeing farming as work happening within a system of relations, and not just human relations (Strand, Arnould, & Press, 2014; Hall & Mogyorody, 2001; Kaltoft, 2001; Lighthall, 1995). Two farmers of this study who consider themselves "strong advocates for organics" remembered the specific event that turned them, in the words of Elmer Laird, "off chemicals." Organic farming in Saskatchewan started with Elmer and his late wife, Gladys, who together, in 1969, transitioned their farm—loamy land bought with government money given WWII veterans—to organic methods, like intercropping using complementary plants (e.g., hemp and pea).[5]

---

5. Elmer showed me alfalfa crops he was using in a rotation because the plants fix nitrogen in the soil, they have long roots that keep moisture there, and the crop gets a high commodity

For Elmer, it was the realization that the chemicals he was spraying on his farmland were making their way into the bodies of family members and those of the entire community. Elmer not only converted his farm, as he put it, "under the framework of ecological principles," but he and Gladys helped start a cooperative local store that would sell only what Elmer describes as "wholesome goods." In Elmer's words, "pesticides threaten all of us." The articulation of healthy farming as farming that works with and for others was extended beyond microscopic features of the land to include interpersonal relationships. A few farmers told me about how they have actively kept their operations small, thereby bolstering the longevity of a traditional rural community. To the urban market gardener "William," farming practice is a means of eliminating involvement in centralized forms of marketing (or getting product to its markets) because he sees that these forms of marketing and managing food relationships are "ideologically at the root of the industrialized food system." Organic production was recognized by many of the farmers I interviewed as circumventing the dominant model for farming by "cutting out the middlemen" in the agricultural chain of production, as one farmer put it.

Biotechnological protest in Saskatchewan ought to be understood in the context of organic farming as protest—as a response to the dominant, productivist[6] strategy and what are perceived by these farmers as its inherent social and environmental impacts (for a scholarly critique of productivism, see McMichael, 2009). "Nina," who managed a certified organic family farm, described her motivations:

> My calculations were never clearly economic but more environmental or social. I see the organic model as in favour of the unstandardized, the unique to every situation, as recognizing the complexities—it just seems to me that the dominant agricultural model misunderstands how extremely complex and interrelated natural ecosystems are. With organics there's a whole other range of thinking: a reconnection of people to communities, to the ecology, and it is production that recognizes our ecological place in ways which are much more sustainable.

Small-scale organic farming is seen, by those farmers I interviewed, as an alternative to the dominant system's treatment of humans and non-human

---

price. When I interviewed him he was also toying with the idea of introducing Leaf Cutter Bees to pollinate the alfalfa in order to secondarily produce seed for use and sale.

6. Productivism is a model of agriculture that describes success by measurable economic productivity and growth. Agriculture organized under productivist logic is characterized by intensive use of land and the growth of staple commodities for export markets, using high technological inputs like biotechnologies.

animals because it promotes a decentralized food system that is responsible to local ways of life and to sustaining environmental health, broadly defined (i.e., including social relationships). Elmer Laird told me that the first female leader of the National Farmers Union had a saying about borrowing the land from our descendants and that this idea of intergenerational responsibility motivated his organic practice: "It really stuck with me," he said.

These Saskatchewan farmers appear to be like organic farmers elsewhere in that they see a link between producing food and producing social structures that look nothing like those given shape under the dominant productivist model of farming (Strand et al., 2014; Hall & Mogyorody, 2001; Barnes & Blevins, 1992; Fiddes, 1991). Productivist farming methods are technologically intensive and aimed at high-yield production of commodity crops destined for export markets. For the farmers of this study, one of the major contentions over biotechnology arises from its attempts to further the industrialization of nature, which project has, for these farmers in their daily experiences, only served to recast agricultural land as capital and place it under "threat," as one farmer put it. Farmers told me that biotechnologies offer a remedy for problems that might be more correctly recognized as inherent instabilities stemming from the technologization of nature under the productivist system. Take what is called monocropping as an example: While growing only one crop can be enormously economically productive, it increases crop vulnerability to disease that then requires technologies like pesticides for farm management (for a recent comprehensive treatment on this see IAASTD, 2009). Also, while proponents of biotechnologies suggest that they increase biodiversity (by inserting novel combinations of genes into crop strains), the farmers I interviewed talked about biodiversity as defined as a diversification of one's crops as well as farm practices in order to provide a holistic or systemic defense against pests, diseases, and market risks.

The farmers I interviewed spoke very critically of agricultural corporations that sell chemical herbicides and pesticides and biotechnologies, in large part because they see them as having eroded rural communities and environmental heath through a singular focus on profit maximization. One farmer, "Dylan," told me in incredible detail that chemical corporations, like Monsanto, gained disproportionate market power through the latter twentieth century, which concentration now allows them to extract monopoly profits. As Dylan put it:

> There are graphs that show grain prices slowly working their way up and fertilizer prices tracking them perfectly. There is so little competition among chemical corporations they have the power to just raise prices at will. And . . .

I think that's what's caused farmers to leave the land, corporations have just squeezed us so hard that we're leaving . . . well . . . we're forced out.

Another farmer put it this way:

The fundamental problem is that while farmers work in a perfect market [perfect meaning competition instead of monopolization], the firms that supply us chemicals aren't competing—they are few in number and they're consolidating or merging and becoming more powerful all the time.

Dylan was in the process of leaving farming and he said to me, with great emotion, "There are real challenges leading to the total disintegration of the rural community." Disintegration is a choice word, evocative of a system coming undone, of farmers becoming disconnected from one another and from the land. Dylan explained this undoing further, saying: "We've lost some of our neighborhood institutions—like the elevators, which got burned five years ago, and the local post office and the store. But we're trying hard to retain a neighborliness and maintain a cohesive network, it's so important."

The erosion of rural communities over the past half-century has happened not just in the material foundation of rural communities—the disappearance of people and then of country stores and schoolhouses—but also in the relationships among farmers. Farmers are especially contemptuous of Monsanto and similar chemical corporations or "life sciences" corporations (those who sell biotechnologies paired with chemicals) for having changed relationships at the local level through the use of seed contracts, which grant farmers limited access to biotech seed systems. Seed contracts prevent farmers by law from saving the seeds for use the next year, thus rendering the seeds, as legal scholar Andre Magnan puts it, "legally sterile" (2004, p. 306). As well, biotech seed systems have shown a tendency to reseed themselves without human intervention and backcross with their hybrid or wild relatives (Friesen, Nelson, & Van Acker, 2003; Andersson & de Vicente, 2010; Huang et al., 2014). If a farmer is caught with biotech material on his or her property, and if that farmer has not signed a contract, he or she is vulnerable to an infringement lawsuit. To the farmers I spoke with, the very presence of biotechnologies on the Prairies has therefore, in the words of "Greg," "pitted farmer against farmer," discouraging cooperation among farmers by creating a climate of suspicion and fear.

From this perspective, the [1998] *Monsanto v. Schmeiser* trial along with the events leading up to it were emblematic of the situation. In 1998, Monsanto's proprietary canola was found growing in Percy Schmeiser's farm fields even though he had not signed the required contract nor paid Monsanto the requisite fee to use their seed technologies. While Schmeiser considered it (and described it) as an unwanted spread of biotech material mingling with his non-biotech seed stock, Monsanto considered this property infringement and sued him. Eventually, in May 2004, four of five Supreme Court justices found Schmeiser guilty of patent infringement though they awarded no damages, which lead Percy Schmeiser to declare (on a now defunct website) a "victory." Monsanto had found out that Roundup Ready canola was growing on Schmeiser's field from one of his neighbors, who reported Schmeiser to the company. This neighbor qua private detective effectively trespassed on Schmeiser's field in order to voluntarily gather samples of canola plants.

During the *Schmeiser* trial, farmer members of The Canadian Canola Growers Association declared themselves on the side of Monsanto because biotechnology was the key to their livelihood (Kinchy, 2012). While these farmers grouped in opposition to Percy Schmeiser, other farmers coalesced in support of him. In the process of this activism, a group of organic farmers formalized themselves into the Saskatchewan Organic Directorate (SOD), an umbrella organization advocating for organic agriculture. I witnessed SOD take shape: farmers sharing stories about biotech "contamination," as they also called it, and worrying together about the future of organic agriculture on the Prairies. Then on February 18, 2002, when the courts were still in the process of deciding whether to uphold Monsanto's right to protect its patent in the appeal ruling of *Schmeiser*, a handful of SOD members announced that they had filed an application to pursue Monsanto in court in [2002] *Hoffman et al. versus Monsanto*, on one level in an attempt to hold corporations responsible for biotechnology contamination. In interviews, however, farmers described the *Hoffman* lawsuit in more expansive terms. One farmer, "Archie," described the lawsuit this way:

> Although very few people are ever able to challenge them [life science corporations], least of all farmers who are labeled as not educated and discounted, there's another picture. There's a whole other range of thinking from that dominant productivist model of agriculture altogether. It's handy to have a label for that other party that's concrete, in this case "Monsanto." But Monsanto of course represents a whole range of interests: corporate interests, and a way of seeing the world as resources to be exploited, a way of looking at people as mobile and exploitable and expendable, and a way of looking at living organisms as manipulable.

Farmers Resist Biotechnology as a "Value System"

Biotechnology resistance on the Canadian Prairie—that happening at the farm field and the courts—is clearly not resistance to the science of recombination per se, as deficit model understandings would have a science communicator believe; instead, dialogical communication with the farmers of this study revealed that their biotechnological resistance can be fitted into a broader framework encapsulating other "counter-publics" of resistance, like environmental or gay rights activists (Warner, 2002). Saskatchewan farmers protest biotechnologies within complex historical relationships of power and inequity and against the backdrop of significant social upheaval that they trace to the presence of life sciences corporations, like Monsanto, in agriculture. Saskatchewan farmers figured themselves in conversations with me as bulwarks against what they see as rampant corporate power. This study's data make clear that Saskatchewan farmers resist biotechnologies not as technological objects, but as material instantiations of the historic inequity and injustice that they feel has shaped the Canadian agricultural system. Nina put it this way:

> Biotechnology is a different value system. The whole value around clean fields, monocultures, maximizing production, not a weed in sight . . . the thousand apples all looking exactly the same way . . . we see this as commentary on the chemicals and the residual toxins rather than seeing it as a commentary of good farming practices. And farmers didn't invent this, it's part of a general cultural bias which permeates the whole agricultural system towards maximizing and standardizing and industrializing everything. Biotechnology represents all of this.

Making sense of resistance to biotechnologies as a product of scientific ignorance, as the deficit model suggests, gives a science communicator little analytical purchase on Canadian Prairie farmers who resist biotechnologies as a way of thinking or, as one farmer put it, "a gold rush mentality." The following statement from "John" makes clear how these technological critiques do not resonate on a terrain that can easily be made sense of using the deficit model:

> I am extremely negative about biotechnologies as a cultural endeavor that's concentrated. . . . I think it's just more of the same idea as the control over natural processes, which will extract a profitability for a very small number. And I think it's completely the wrong view of ourselves within the web of life.

UNETHICAL RESULTS FROM USING A DEFICIT MODEL

My interviews also revealed a kind of politics to the deficit model framing of Saskatchewan farmers as anti-science and ignorant: Deeming citizens technically illiterate appears to motivate and justify the continued exclusion of particular actors from the shaping of agricultural agendas. Almost every farmer I spoke with expressed feelings of exclusion from participating in decision making on biotechnologies. They told me story after story about failed attempts to engage in the formal political channels of decision making, such as letters sent to the Agricultural Minister of Saskatchewan (then Lyle Vanclief) for which they got no response. Furthermore, they highlight this arguably democratic issue as being at the root of their resistance, both within and without of the courts. In the words of one farmer: "Despite ongoing lobbying efforts to get full public hearings on the issue of GE wheat, to date there has been no action, nobody is listening to us, and we hope the [Hoffman] lawsuit can do something to change this." Incidentally, farmer resistance around biotech wheat in particular prevented its commercialization (more on this below).

This study's data also suggest that the exclusion of particular voices from biotechnological decision making is happening not just within formal political channels of participation but also in the setting of regulatory and agricultural research priorities. Nina described for me how in her mind, "a biotechnological frame of thinking" dominates the current agricultural system, including expert decision making, the funding of public research, and the direction of farm-level support:

> The policy and international trade agreements and expertise are all geared towards looking at the industrial segment. They [government] are being so aggressive in support of GM foods and biotechnology research.

The biotechnology industry in particular is seen to be benefitting from, in the words of Dylan, "huge government investments in biotechnology," which is rationalized as a consequence of the perceived importance, among agro-economic experts, of biotechnology to the national economy: "[The Canadian government] sees biotechnology as Canada's niche in the global economy." This claim appears to be supported by the numbers. The federal government, via Agriculture and Agri-Food Canada, invested something over $8.6 million in organic research between 2010 and 2013. By contrast, the Genomics Research and Development Initiative, the funding program for genomics research in federal laboratories, received $86.3 million in funding from Agriculture and Agri-Food Canada between 1999 and 2013 (Genome Canada, 2013; Government of Canada, 1999).

Farmers make sense of the dominance of, in John's words, "that corporate model" as a consequence not only of formal ties between government and industry (for instance, early governmental investment into biotechnology), but also as part of "a long-standing cultural bias for the industrial way of farming" and toward productivist farming strategies. Farmers see Agriculture and Agri-Food Canada as creating problems that can then be turned into commercial opportunities. Participants spoke about how the Canadian government's agricultural experts were "pushing" biotechnologies as environmentally benign and even an answer to problems in agriculture, like soil erosion, that can instead be seen as consequences of technologization. Nina told me that "the biotechnological project" is a kind of conceptual framework that engenders technological solutions to what are not at root technical problems, while at the same time it forecloses the future by diverting resources from alternative natures (and human natures) that might otherwise be fostered. Canadian decision makers, I was told by many farmers, are a kind of narrow-minded pawn in a larger system, where agricultural experts have accepted the corporate problem definitions for the food system, prioritizing research and development that takes up high technologies, like biotechnologies, as strategies for problems like global food insecurity—strategies that these farmers see as primarily advantaging large agri-food corporations. Nina put it this way:

> Monsanto really represents a way of seeing and a way of organizing the political domain such that fewer and fewer actors determine what life, the environment, and economy will be like without reference to people's preferences, their citizenry, and their democratic rights.

According to these Saskatchewan farmers, the problem with biotechnologies is not so much the existence of particular technological forms, or even the existence of the values they represent, but that these values dominate the agricultural agenda at the expense of alternative ways of organizing life. For Saskatchewan farmers involved in legal battles against Monsanto, the corporation is therefore a symbol for a hegemonic *way of seeing*, and their legal activism is on one level an attempt to give voice to alternative ways of envisioning rural life and distributing justice across the agricultural system (i.e., beyond the interests of large corporations).

Farmers also described to me how the interests of large life sciences corporations are served by the social and political context of scientific facts used in biotechnology regulation. In interviews, farmers described their concerns over the limitations of the facts being used in regulatory decision making, as well as the context in which these regulatory facts are derived—corporate, not public, laboratories. Farmers described in detail the limitations of the current working conception of risk within biotechnology regulation: a reductionist framing of biotechnological risk that does not account for the laboratory process of recombination as possibly creating yet unknown or even unknowable hazards (Bronson, 2014b; Wynne & Bronson, 2016). Indeed, in official documents, biotechnology regulators seem to describe risk as a technical problem for the laboratory rather than an overall stance toward possible outcomes from technologies.

The farmers I interviewed seemed concerned that this is a limited conception of risk in both time and space. Farmers considered the possibility that biotech seed systems may present longer-term effects that would go unnoticed within a reductionist scientific approach. As well, these farmers felt that there are hazards (including the possibility of biotech seed and pollen spread) that fail to appear within the current risk-management approach because they present themselves outside of the confines of a laboratory. Most of all, these farmers expressed concern that there is no process of public deliberation on these decisions about how risk is being defined within the regulatory context, thus rendering invisible the fact that these are decisions taken by people in particular social and political contexts (e.g., long-standing governmental investment into biotechnology.

Conclusion

So what insights might practitioners or science communication scholars draw from this case study, especially those wanting to consider the ethics of science communication in scholarship and practice? This case raises a clear ethical dimension to the deficit model's use. The deficit model not only fails to adequately capture the range and complexity of social, political, economic, and environmental concerns that inform western Canadian farmers' critiques of biotechnologies, but also takes on a role in a long-standing network of power relations between farmers, corporations, and governance actors. The deficit model appears to motivate and justify the exclusion of key actors from the setting of democratic priorities and thus perpetuates what are perceived to be long-standing injustices in the food system, such as inequitable relationships of power between farmers and agri-business (for academic work on power inequities, see Friedmann, 2009; McMichael, 2009; Clapp, 2012).

This case study, then, suggests the utility as well as the normative value of a contextual approach to science communication. Taking such an approach allowed me to recognize that at the epicenter of western Canadian farmer resistance to biotech is a frustration about the lack of openness in biotechnological governance. Farmers expressed concern over a lack of public deliberation in policy-making processes, but also a lack of transparency to the political and corporate context that frames the science used in biotechnology policy making. Farmers said that a lack of public deliberation about regulatory decisions—an exclusion justified by deficit model assumptions, including decisions about the choice of scientific frameworks—concealed the fact that value-based decisions are being made by people in particular social and political contexts. Faced with a political system that appears to be justifying the exclusion of public concerns by conveniently labeling them technical ignorance, western Canadian farmers, like farmers elsewhere (Fitting, 2010), were using the Canadian courts as a site for activism. The statement to the court in *Hoffman* asks for a legal declaration of biotechnology as a "development" in environmental law where

> a successful declaration that the testing and development of GMOs was a "development" within the meaning of the EA would operate to modify behavior because the Defendants can be enjoined if they should attempt to introduce future GM crops without ministerial approval. This would compel the Defendants to submit their engineered gene to a public environmental scrutiny rather than the behind closed doors approach they have been allowed to use with the federal government's regulatory bodies. (Memorandum of Appeal [2002] para. 17)

Legal contestations between farmers and life sciences corporations are thus a last-resort platform for bringing forward democratic disputes around biotechnologies in the absence of other communication with decision makers. Rather than resisting biotechnologies because of fear or technical ignorance, Saskatchewan farmers readily identify, evaluate, and act on the social consequences of biotechnologies that are visible from the contexts of their own lives. This group of Canadian farmers is attempting to articulate an alternative agricultural science (e.g., assessments of biotech risk), in which we can recognize a history of ideas and political associations that they see as having shaped the Canadian Prairie and its social and cultural environment in negative ways.

From within the deficit model, however, Prairie farmer biotechnological critiques are divorced from their historical context and are rendered expressions of public illiteracy. Misunderstood as public misunderstanding of science, the biotech dispute continues to bring enormous pressure to bear on the biotech industry and on policy makers. In 2004, public pressure—from environmental groups, consumer groups, and largely from a highly organized grain farming sector—forced Monsanto to shelve its Roundup Ready wheat that was set for commercialization. It is possible that others of Monsanto's products (notably, alfalfa) could go the same way: In 2010, a Member of Parliament from British Columbia introduced Bill C-474 calling for an amendment to the Seeds Regulations that would require an analysis of potential harm to export markets before the commercialization of any new biotech seed system. A writer on behalf of the Canadian Organic Growers—an advocacy organization for organic farmers in Canada—voiced their support for this bill, citing a deeply flawed national regulatory protocol for dealing with public concern.

The present case study joins with other in-depth research on biotechnological engagement that calls for new ways of conceptualizing "the problem" of technology resistance (e.g., Kinchy, 2012; Wynne, 2008; Priest, 2000). What gets described, as Steve Paikin did, as conflicts between science versus values is often better understood as a rejection of the values underlying the taken-for-granted scientific consensus among experts. The farmers presented in this chapter raise a challenge to science insofar as science already embodies and furthers political and normative social commitments (Wynne, 1982); such scientific critiques are easier to understand if one remains open to viewing scientific knowledge as pursued and obtained within particular sociopolitical and ideological contexts, rather than as fixed knowledge from which any difference signals ignorance.

References

Andersson, M. S., & Vicente, M. C. de. (2010). *Gene flow between crops and their wild relatives.* Baltimore, MD: Johns Hopkins University Press.

Barnes, D., & Blevins, A. (1992). Farm structure and the economic well-being of nonmetropolitan counties. *Rural Sociology, 57*(3), 333–346.

Bijker, W., Bal, R., & Hendriks, R. (2009). The paradox of scientific authority, the role of scientific advice in democracies. Cambridge, MA: MIT Press.

Bora, A. (2010). Technoscientific normativity and the "iron cage" of law. *Science, Technology, & Human Values, 35*(1), 3–28.

Bronson, K. (2009). What we talk about when we talk about biotechnology. *Politics and Culture 2* (Special issue on food sovereignty). Retrieved from http://maxhaiven.com/2009/04/07/foodsovereignty-a-special-issue-of-politics-and-culture-edited-by-scott-stoneman-and-max-haiven/

Bronson, K. (2014a). Reflecting on the science in science communication. *Canadian Journal of Communication, 39*(3), 523–537.

Bronson, K. (2014b). The shaping of science in biotechnology conflicts. [Review of the book *Seeds, Science, and Struggle* by A. Kinchy]. *Science as Culture, 23*(4), 580–584.

Brossard, D., & Lewenstein, B. (2010). A critical appraisal of models of public understanding of science. In L. Kahlor & P. A. Stout (Eds.), *Communicating Science: New agendas in communication* (pp. 11–39). New York, NY: Routledge.

Byrne, P., Namuth, D., Harrington, J., Ward, S., Lee, D., & Hain, P. (2002). Increasing public understanding of transgenic crops through the world wide web. *Public Understanding of Science, 11*(3), 293–304.

Clapp, J. (2012). *Food.* Cambridge, UK: Polity.

Cook, G. (2004). *Genetically modified language: The discourse of the GM debate.* New York, NY: Routledge.

Cook, G., & Robbins, P. (2002). *The presentation of GM crop research to non-specialists: A case study.* Reading, UK: University of Reading.

Davies, S. (2009). Learning to engage; engaging to learn; the purpose of informal science-public dialogue. In R. Holliman, E. Whiteleg, E. Scanlan, S. Smidt, & J. Thomas (Eds.), *Investigating science communication in the information age*. Oxford, UK: Oxford University Press.

Durant, J., Evans, G., & Thomas, G. (1989). The public understanding of science. *Nature 340*, 11–14.

Fiddes, N. (1991). *Meat: A natural symbol*. London, UK: Routledge.

Fitting, E. (2010). *The struggle for maize*. Durham, NC: Duke University Press.

Friedmann, H. (2009). Feeding the empire: The pathologies of globalized agriculture. *Socialist Register, 41*, 124–143.

Friesen, L., Nelson, A., & Van Acker, R. (2003). Evidence of contamination of pedigreed canola (*Brassica napus*) seedlots in Western Canada with genetically engineered herbicide resistance traits. *Agronomy Journal, 95*(5), 1342–1347.

Genome Canada. (2013). The opportunity for agri-food genomics in Canada, ours for the making. Retrieved from http://www.genomecanada.ca/sites/genomecanada/files/sector/Agri_Food_EN.pdf

Government of Canada. (1999). Organic agricultural standards development history. Retrieved from http://www.tpsgc-pwgsc.gc.ca/ongc-cgsb/programme-program/normes-standards/comm/32-20-agriculture-eng.html#a3

Government of Canada. (2013). Organic value chain roundtable strategic plan. Retrieved from http://www.agr.gc.ca/eng/industry-markets-and-trade/value-chain-roundtables/organics/strategic-plan/?id=1399477456800

Gregory, J., & Locke, S. J. (2008). The evolution of "public understanding of science": Public engagement as a tool of science policy in the UK. *Sociology Compass 2*(4), 1252–1265.

Gregory, J., & Miller, S. (1998). *Science in public: Communication, culture, and credibility*. Cambridge, MA: Perseus.

Hall, A., & Mogyorody, V. (2001). Organic farmers in Ontario: An examination of the conventionalization argument. *Sociologia Ruralis 41*(4), 399–422.

Haraway, D. (1995). Situated knowledges: The science question in feminism and the privilege of partial perspective. *Feminist Studies, 14*(3), 575–599.

Hargrove, T. (2001). 8 months prison recommended for Jose Bove over destruction of GM rice plants. Institute for Agriculture and Trade Policy. Retrieved from http://www.iatp.org/news/8-months-prison-recommended-for-jose-bove-over-destruction-of-gm-rice-plants#sthash.V3xeMzyI.dpuf

Hart, S., & Nisbet, E. (2012). Boomerang effects in science communication: How motivated reasoning and identity cues amplify opinion polarization about climate mitigation. *Communication Research, 39*(6), 701–723.

Huang, W.-K., Peng, H., Wang, G.-F., Cui, J.-K., Zhu, L.-F., Long, H.-B., & Peng, D.-L. (2014). Assessment of gene flow from glyphosate-resistant transgenic soybean to conventional soybean in China. *Acta Physiologiae Plantarum, 36*(7), 1637–1647.

International Assessment of Agricultural Knowledge, Science, Technology for Development (IAASTD). (2009). Agriculture at a crossroads: Synthesis report for the UN Environment Program. Retrieved from http://www.unep.org/dewa/agassessment/reports/IAASTD/EN/Agriculture%20at%20a%20Crossroads_Synthesis%20Report%20(English).pdf

Irwin, A., & Michael, M. (Eds.). (2003). *Science, social theory and public knowledge*. Philadelphia, PA: Open University Press.

Irwin, A., & Wynne, B. (1996). *Misunderstanding science? The public reconstruction of science and technology*. Cambridge, UK: Cambridge University Press.

Kaltoft, P. (2001). Organic farming in late modernity: At the frontier of modernity or opposing modernity? *Sociologia Ruralis, 41*(2), 146–158.

Kinchy, A. (2012). *Seeds, science, and struggle: The global politics of transgenic crops.* Cambridge, MA: MIT.

Knight, A. (2009). Perceptions, knowledge and ethical concerns with GM foods and GM process. *Public Understanding of Science, 18*(2), 177–188.

Lighthall, D. (1995). Farm structure and chemical use in the corn belt. *Rural Sociology, 60*(3), 505–520.

Lynas, Mark. (2015, April 24). How I got converted to GMO food. *The New York Times.* Retrieved from http://www.nytimes.com/2015/04/25/opinion/sunday/how-i-got-converted -to-gmo-food.html?_r=0

Magnan, A. (2004). Social and political implications of genetically modified crops in Saskatchewan. *Prairie Forum 29*(2), 301–316.

McMichael, P. (2009). A food regime genealogy. *Journal of Peasant Studies, 36*(1), 139–169.

Montpetit, E., & Rouillard, C. (2008). Culture and the democratization of risk management: The widening biotechnology gap between Canada and France. *Administration and Society, 39,* 907–930.

Nisbet, M., & Scheufele, D. (2009). What's next for science communication? Promising directions and lingering distractions. *American Journal of Botany, 96*(10), 1767–1778.

Priest, S. (2000). US public opinion divided over biotechnology? *Nature, 18,* 939–942.

Prno, J., & Slocombe, S. (2012). Exploring the origins of "social license to operate in the mining sector": Perspectives from governance and sustainability theories. *Resources Policy, 37*(3), 346–357.

Royal Society. (1985). The public understanding of science. Report of an ad hoc group chaired by W. F. Bodmer. London, UK: Royal Society.

Shields, R. (2008). Hope and fear in biotechnology: The space-times of regulatory affect. *Space and Culture, 11*(2), 125–141.

Sismondo, S. (2004). *An introduction to science and technology studies.* Oxford, UK: Blackwell.

Strand, K., Arnould E., & Press, M. (2014). Tillage practices and identity formation in high plains farming. *Journal of Material Culture, 19*(4), 355–373.

SwissInfo. (2007). Two killed in shoot-out at Syngenta GM farm. Retrieved from http://www.swissinfo.ch/eng/Two_killed_in_shootout_at_Syngenta_GM_farm.html?cid= 6208040

Turner, S. (2008). School science and its controversies; or, whatever happened to scientific literacy? *Public Understanding of Science, 17*(1), 55–72.

Warner, M. (2002). *Publics and counterpublics.* Cambridge, MA: MIT Press.

Wynne, B. (1982). *Rationality and ritual: Participation and exclusion in nuclear decision making.* London, UK: EarthScan.

Wynne, B. (1991). Knowledges in context. *Science, Technology and Human Values, 16,* 111–121.

Wynne, B. (2008). Public participation in science and technology: Performing and obscuring a political-conceptual category mistake. *East Asian Science, Technology and Society: An International Journal, 1*(1), 99–110.

Wynne, B., & Bronson, I. (2016, January). *Scientific standards as politics by other means: The case of environmental risk assessment for GM crops.* Paper presented at the Making Standards Workshop, Dalhousie University, Halifax, Nova Scotia, Canada.

Yearly, S. (2005). *Making sense of science: Understanding the social study of science.* London, UK: Sage.

13

Science Communication Ethics:
A Reflexive View

ALAIN LÉTOURNEAU

This chapter is situated in practical philosophy (see Létourneau, 2014), which means a centering on human action that has to be studied carefully. Starting with the Greeks through Kant, philosophical pragmatism, and recent analytical efforts, practical philosophy includes philosophy of law, political philosophy, and the different branches of ethics, especially different kinds of applied ethics, but we also need to include in it the actual study of organized practices—because normative considerations, which are classically called for when we discuss law, ethics, and politics, would not make sense without taking into account the conditions, contexts, and internal structure of action practices. This implies a radical openness to the numerous disciplines that are interested in human action and to social science approaches based on empirical data—practical philosophy cannot therefore be limited to a discussion of the normative, of what "we ought" to do. Such a perspective aims to set itself some limits; here we will focus on practices of communicating science, quite a large domain as we will indicate further on.

Classical philosophical pragmatism (especially Peirce, James, Dewey, and Mead) joins here with hermeneutics (Dilthey and Ricoeur) in a frame of thought inspired by critical theory (Habermas, Apel, Honneth). This already implies that theories are seen not as pure representational images of reality, but as tools constructed with certain aims, with certain questions in mind; they also are not independent, as research enterprises, of value orientations (Weber, 1992). Theorizing is always taking place in a given context, which happens because there is a valuation of certain elements as tentative answers to specific questions (Dewey, 1939). Argumentation and rhetoric can be distinguished but united—argumentation standing for analysis or argumentation schemes and rhetoric mostly interested in how communication affects people, whether it be by the use of tropes, figures, and other appeals to perception and emotion, while taking into account the roles of the audience and "publics" in constructing a message (Tindale, 1999; Dewey, 1927; Perelman and Olberechts-Tyteca, 1988). I would also like to introduce a pragmatist notion of "meanings-in-use"; it is one thing to consider what philosophers might think or theorize about ethics, but at least in some cases, it is another thing to look at what people actually mean and do when they use the word ethics. For me it is important to be clear about this when we discuss ethics, because there are social requirements and expectations about it that might either trump or play a part in our efforts.

Ethics can be seen as something profoundly reflexive and critical; for me a dialogical-deliberative dimension is forcibly required and is a constitutive part of it (Létourneau, 2014; but also see Rawls, 1993; Habermas, 1982, 1997). Ethics needs the reflexive perspective: This means that it renders the value and normative claims held by actors and speakers explicit, as they emerge in actual social practices and discourses. Reflexivity here also implies the pragmatist key idea according to which social norms and values are ways for actors to try to give an answer to problematic situations (Dewey, 1939). This amounts to saying that ethics can be seen as a part of practical philosophy; it is produced in interaction with communication and other social and natural sciences.

This chapter seeks to clarify what "ethics of science communication" really means. This implies first to look at what we mean by ethics, as well as by science communication. To provide some background on the first point, we refer to a previous study on the meanings-in-use of the word "ethics." It turns out that a plurality of meanings of the word "ethics" can have a valid claim here. To better understand what science communication is, we will then develop a few of the practice contexts that can characterize it. A plurality of readings of science communication will become apparent on the basis of these different readings of "ethics," especially when we take into account the plurality of potential "receivers" of science communication. The different types of science communication practice that exist could be interpreted differently, provided we understand ethics in one way or the other. One specific science communication example will be looked at to illustrate one type of that kind of communication, raising one type of reading of what "ethics" means; the case study will not illustrate all possible ethical understandings. We will discuss a piece originally published on the website of a research center in plant science based in Québec, Canada, which is called Centre Sève. In science communication as elsewhere, normative and axiological issues will arise; some discussion will be made on that point, a question that is certainly difficult and as complex as the different organizational settings can be.

Ethics: Meanings-in-Use and Reflexive Turn

A first step is to characterize more fully our understanding of ethics. A reflexive ethic is different from a behavioristic or purely normative approach; we see ethics as a dialogic effort, which means for us to take into account the different involvement of actors on issues, while trying to facilitate a process of thinking together beyond entrenched positions (Buber, 1922; Legault, 1998; Isaacs, 1999; Létourneau, 2012). This dialogue has to do with the axiological and normative dimensions of human action in specific contexts, with the aim of identifying and then, if possible, helping people find their own solutions to these difficulties. This does not deny problems that have to do with power distribution, profound disagreement, or resource allocation, but it is certainly related to an ideal of democratic life (Dewey, 1927). If and when issues are effectively treated by people who aim to "walk the talk," it can be both meaningful and operative, at least according to some actors.[1] Similar results have been also documented in other organizational settings by referring to reframing procedures, among other similar processes (Putnam & Holmer, 1992). Later in this chapter, these processes will be illustrated by a case study of how a particular communication product was developed by a team of scientists.

But we cannot just posit an ideal model without looking at what people think and actually do. It might be useful to recall a previous analytical survey of the uses of the word "ethics" in three of the main French newspapers published in Québec Province in a period of three years (from 2000 to 2003; Létourneau, 2005). The aim was to characterize how ethics was understood in the uses of the press, even though newspaper articles almost never take the time to define something like "ethics"—in fact they surmise that there is a common or general understanding of this word, with which they can operate. It is possible to reconstruct the basic meaning that they give to the word by examining the use of it that the articles display.

In that research on the word "ethics," the uses of the word were grouped under six categories: (1) behavioristic, where basically we are discussing ethics

---

1. I would refer people here to experiences being implemented in what is now being called in Canada Organizational Ethics. See for instance RÉOQ, *Regroupement en éthique organisationnelle du Québec;* their website is at http://www.reoq.ca/

in terms of good and bad behaviors, with formulations similar to "to be ethical" or "unethical"; (2) normative, where ethics is discussed in terms of the presence or absence of a clear set of rules that are most of the time deemed either required or insufficient; (3) expert, where it is associated with specialized actors, the word "ethics" being then given with reference to advisors, ethics officers, committees, or commissions devoted to ethics; (4) value-oriented, where ethics is associated with mission values, and also good values, for instance a "family" ethic or a "citizen's" ethic; (5) reflexive, where ethics is considered more philosophically as a critical reflection on norms and values enacted, a use almost always associated with scholars of philosophy or ethics; and, finally, (6) references to ethics as financial, for instance ethical funds, also referred to as ethical investments.

This project characterized *how* written media *talk* when they use the word ethic(s), without focusing on specific themes like corruption or other disputable behavior. A striking result was that globally speaking, meanings 1, 2, and 3 are dominating the corpus with minor quantitative differences related to different elements. More than that, these three meanings tend to converge and coalesce in a perspective according to which ethics aims at good behaviors in given groups of actors by using sets of specific norms that are interpreted by experts called "ethicians" or "ethics officers." The question of the kinds of behaviors discussed (and the ones that are not) would, of course, also be another meaningful study. The question of whether our results would also be obtained with a different, larger corpus (for instance, in settings in the United States, France, or English-Speaking Canada) is an interesting one that cannot be addressed here. But there is no important reason to suppose prima facie that results would not be similar elsewhere, at least in the Western world.[2]

These meanings-in-use command very different views and understandings of the task of ethics. It is one thing to characterize the task of ethics as having to teach how to follow norms that would precisely orient action, for instance as in so-called codes of ethics or professional codes of conduct; it is another to help people understand how values orient their choices, and how they have to make those choices more conscious and deliberate, to the point of being able to justify them publicly while taking into account action possibilities and the different consequences of different courses of action (Legault, 1998). Said otherwise, a value-oriented meaning combined with a reflective approach (meanings 4 and 5) can be effectively contrasted with meanings 1 to 3 combined. While these understandings can sometimes overlap, are not necessarily exclusive, and can be taken into account together, differences of

2. For quantitative details see Létourneau (2005).

emphasis do exist and produce very different orientations, procedures, ends pursued, and also consequences. It cannot be the same thing to think that ethics mostly means to have a good and precise set of rules guiding behavior, and to think that it has mainly to do with helping people becoming more conscious of their real value involvement on issues. Yet both orientations are completely possible and actually exist (Létourneau, 2007).

Of course people can pursue an ideal of being able to discriminate between good and bad behavior, but in some circumstances, such a distinction is quite difficult, if not almost impossible in very hard cases. One easy way to deal with this difficulty is to fall back on relativism, asserting that all is equal, positing that the options are relative to points of view and therefore subjective opinions. But this is not really an option; the tension and sometimes opposition between choices that actors actually have to make show the importance of these values, at least for the people defending them. In ethical life, the choice is not between good and bad, but between different, sometimes incompatible goods, aims, values. For me, realism implies recognizing conflicts between values and even between norms in given situations, when they exist. In complex contemporary societies, we have more and more norms that aim at different goals, and important values can and do come into conflict. When the norm or the value is clear, and we do not find a conflict between a plurality of elements orienting us in differing directions, the need for reflection might be less important, but this is not always the case.

This language-in-use referring to ethical standpoints or values also expresses a social demand for ethics, which is quite important nowadays. And we usually tend to think that a good way to meet this social demand and further "ethicality" would be to have good and precise norms telling us what to do. But surely we can see that in science communication the task of producing a specific set of norms would be very difficult, if only because of the plurality of practices that exist. The situation of the freelance communication professional is not the same as the situation of a scientist interviewed by a newspaper, say for the big new grant just obtained that is going to create jobs nearby. To think it would be easy to fabricate precise rules covering all those different situations might be a mistake.

Normally, professional practice comes with some ethical code, most of the time understood as a deontological code of professional conduct: doctors, lawyers, even researchers have such codes. In the French-speaking world, the term "Code de déontologie" is used to speak of these.[3] A proper set of rules

---

3. A few examples among many: http://www.asha.org/Code-of-Ethics/, and also http://www.ashg.org/pages/about_ethics.shtml

or a "deontology" presupposes a professional setting that is quite well documented, established and with very specific problems that surface and give regulators reasons to develop rules that can apply to these problems. All this presupposes a history, a common set of practices known and documented, from which we can say that such and such specific acts are or would be derogatory or mandatory. Deontology goes with professional settings, involving in particular, at least in Canada, what is called trustees, that is, officers to whom the public can express complaints about behaviors, and who will be authorized to inquire and verify what the practice was, with the important capacity of providing some form of punishment that might in some cases be purely symbolic or not, for instance suspension of the right of practice for a given period of time. A practice that is not well established and documented is very far from any effort at deontology in that sense. The total absence of centralization in such a case obviously forbids the existence of punishments of any kind, at least in the current setting of things. Furthermore, we should remember that with the plurality of media in use in science communication, entailing a variety of requirements and specificities, specific challenges are also to be considered. In any case, such a normative/compliance model with definite sanctions might not be needed in the first place.

In science communication, aiming at good behaviors could be to focus on adequacy of the meaning conveyed, on the ability to deliver according to the mandate given, in a professional way. Aiming at the norms would be trying to look for the norms of science communication in existence, and if they do not yet exist, determine a set of norms and pursue afterward compliance of the practitioners to these sets of norms. The third meaning would imply to have a new kind of ethicist, specializing in science communication: This person would need to be recognized as a specialist, and we would work at establishing that new expert practice domain, probably with the help of a valid and complex set of norms, indicating what to do and not to do in science communication. By contrast, science communication understood in terms of values, combining here this fourth meaning with reflexivity, would instead inspire an ethic aimed at reflecting the actual values of science practices, explicating the values and norms involved in given situations, and put them ahead in the discussion. Communicating science would then aim at communicating not only data or information, but science itself, as an attitude toward inquiry, as the value of the research itself, as openness to the richness of what is to be known and its usefulness, with the requirement also of working in communities, of researchers, of sciences, and of people using science and its results.

As was discussed earlier, ethics can be understood as reflecting on values, rendering them explicit, basically expressing them, when possible with the stakeholders involved. If science communication is considered as building knowledge along with attitudes, it could also aim at illustrating the values involved in scientific research, as a way to sensitize people on its importance, to convince them of its value and of the value of the research. By furthering its development and its grounding in social practices, the outcome might be quite different than a pure information-transfer perspective, and, at least in some cases, it might be more appealing and more efficient also in terms of information. The case study later on in this article gives us an example of one approach close to the fourth meaning explained before.

Science Communication: Different Roles

Science communication can be looked at from an organizational and inter-organizational point of view, those social settings and relationships form-ing the material and social conditions of the possibility of that kind of social practice. Communication does not happen in a vacuum; individuals are in social settings that enable them to act. Science is built in interactions both with the object-domain of investigation and with a plurality of groups: the relevant scientific community or communities, science publics and users. Discourses held by singular people can be better understood if we put them in their organizational contexts: this particular lab, this specific government bureau, inside their relevant networks of other organizational actors. And the problem we have in "science communication ethics" obtains here both a new level of clarity and some new difficulties.

Formulating a few questions will illustrate the situation. Are we looking at *journalists* who have to cover science events? In that case, what is their organi-zational context: Do they act as freelancers (being somehow their own bosses, but getting hired by outside organizations for specific mandates) or do they work for a specific news business, which has characteristics in terms of aims, means, publics, advertisement support, and distribution? They could also be working for public organizations, for instance NPR, the CBC, BBC, and the like. Do we have here classical media (newspapers, magazines, radio or tele-vision programming) or are we in the so-called new media sphere, existing mainly on the web? Aren't there specific values that come into play in these different social/organizational settings? The professional is quite different in one or another of these situations. To work in a traditional newspaper is not the same as working in a blog; to use email is not the same as to use Facebook or Twitter; and these are not the same as other currently existing media, or those yet to be created. With specific platforms and media come specific pos-sibilities and constraints, the presence or not of given protections, and so on.

Another group of actors is also obviously concerned with "science communication": *professional communicators* working for organizations of many different kinds. There are certainly firms for which the main order of business is science or is closely connected to scientific methods, technologies, and approaches. People can be professional communicators in science for government agencies, for instance the EPA or other public organizations. There are, of course, also professional communicators who act as members of social movements or of advocacy organizations, neither government nor for profit, having their own specificities in many cases. Some communication professionals can also get hired by networks of researchers regrouped in funded clusters of excellence and the like, and, of course, think tanks might also need them.

But there is also another distinct possibility: communication by the *scientists* themselves, expressing their current work, including new discoveries or other dimensions; this can also happen in the case of a network of scientific researchers who decide to do the work themselves. The case study discussed below falls into that last category. If only for funding reasons, many research networks are acting that way, taking charge of their own communication. Typically then, a scientist will "specialize" in communication as a kind of new mission. Scientists can also communicate inside social movements, for instance if they are involved in the protection of biodiversity at the local level. Of course, this is not an easy path for scientists, and caution may be required in the face of certain causes.

In the context of the economic restructuring by the old media in the face of the emergence of new communication platforms, organized groups of researchers and individual scientists have more and more opportunities to write in the form of blogs, social media, and so on. With these new contexts, new demands and responsibilities arise that are not completely codified because of their relative novelty. Our three different kinds of science communication actors (journalists, communicators, and scientists) can all act on these new media or platforms. They can all have a lighter or a stronger professional link with employers, more or less autonomy, and better or worse working conditions.

These different forms of science communication—for the public, private, or civil society; professional communication for private or public enterprises focusing on science, or for science organizations whether they are public or private; and researchers communicating as individuals or as part of research groups, or in some cases also within social movements—can hardly be equated on the professional level. For each of these concrete settings, with all the variety that can characterize them otherwise in terms of action domains (biological agriculture, hydrology, branches of physics, biotechnology, engineering, earth sciences, environment, and so on) and in terms of media platforms (classical or new media or both), determining norms and good behaviors will probably have to be tailor-made to situations and contexts that vary considerably.

Instead of looking for norms and behaviors, it might be more useful for the practitioners to ask themselves: What is really important in the communication I am responsible for? What has value for the researchers, the actual or future users of the science, the different groups involved, the general public? What is meaningful and really worth emphasizing in this communication act that is expected? And, of course, also, can I identify the normative constraints that characterize my situation, understood as a complex network of organized actors that have their own requirements? Deadlines, deliverables, quantities and quality of material required and expected, level of professional freedom recognized, value of the imaging displays possible, presentation rules, partners available with whom I might work to perfect the product are all elements that vary enormously from one context to the next. If we want, then, to look more closely at science communication ethics, the next logical step would probably be to document and reflect on the basis of these different sets of practices.

Case Study

A careful reading (without judgmental posturing) of a communication document produced in and by agents in an organization, to present/represent officially that organization, might reveal how values and missions have to do with groups in interaction, aiming to fulfill specific needs. As a case where communication is taken charge of by scientists, e.g., the members of a research network, and not by putting in play professional communicators, here I will introduce and discuss a communication piece created by Centre Sève, an important Québec network of research in plant biology, with which I have a close relationship.[4] The Centre Sève started in 2004 and it is now in its second consecutive six-year grant. It is an infrastructure for research that regroups more than 55 plant science researchers from five different universities (plus one philosopher and two social scientists) to form a network of plant science researchers who have repeatedly obtained provincial, national, and international grants.

---

4. In the interest of transparency, I should clarify that I am currently part of the board of directors of that network, with which I have been collaborating since 2004.

Except at its very beginning, the Centre has not been covered very much over the years by the media, even though it is a very productive and successful research structure. Of course some specific members of the Centre or sub-units of it have been regularly interviewed by media, but the network infra-structure funding group that is sustained by public funds is basically ignored by the public. The Centre as such is not a source of very "hot" news for magazines or newspapers that are aimed at the general public. People operating the Centre never deemed it necessary to actually give a specific mandate to any particular communication professional; as is often the case in science venues, communication seems to be seen as a necessity that is only secondary compared to the main business of research. The document below, which this discussion will analyze in detail, has been displayed on the Centre website since 2011. This is an English version provided by the Centre, alongside a French version also available on their website at www.centreseve.org.

Le Centre SÈVE:
The main mission of the Centre SÈVE is to acquire, disseminate and transfer new knowledge, useful tools and innovative technologies to improve plant productivity and the economic value of agricultural production within a so-cial context that values the sustainability of ecosystems and food security.

The second most important mission of Centre SÈVE is to contribute to the training of highly qualified personnel necessary for the development of Quebec agriculture. The ultimate goals of the Centre SÈVE are to exploit, develop and rationally diversify non-forestry plant resources.

In order to achieve this mission, the Centre SÈVE supports fundamental and applied research at different levels: from molecules to ecosystems; from fundamental knowledge to technology transfer and the production of tools that support decision making. The Centre SÈVE also stands out for the extent of its expertise and network as well as its interdisciplinary initiatives.

More specifically, the Centre SÈVE's objectives are:

— the undertaking of scientific research that aims to develop knowledge and tools to increase plant productivity while taking into account various environmental and social concerns;
— integrating the expertise and infrastructure of 62 researchers from six universities and two government institutions;
— knowledge transfer from Centre SÈVE scientists to stakeholders, in-cluding students at all levels, and more widely to the general public, during training workshops, seminars, conferences, science and tech-nology meetings and public forums;
— training of qualified personnel to meet future needs in the areas of plant science and agriculture, offered in a multidisciplinary and in-novative setting that promotes high quality training and research;

— communication between researchers in plant sciences and researchers
   in social and environmental sciences;
— establishing national and international collaborations to provide new
   research and training opportunities for our researchers and students;
— providing a platform to highlight the expertise of Québec researchers
   both nationally and internationally.

This text is a general statement of missions that attributes value to specific actions (Dewey, 1939); in that sense, the document manifests the public
self-understanding of the Centre as an expression of its purposes, first the
pursuit of science itself in close relationship with science users. Thus it is a
relevant document according to the fourth meaning-in-use previously discussed (value orientation). It is obviously intended to reach a broad public,
even though some knowledge of the relevant research fields is needed to fully
grasp what is written here. From the standpoint of practical philosophy, we
have here a series of action programs that are somehow coordinated in the
text, themselves being closely related to values, not to norms. A number of
subsets of organized research groups are indicated here, even though they
are not explicitly mentioned in each case. All the paragraphs indicate actual
groups.

In this document, the Centre explains its two main missions and situates
its support of research initiatives that are partly listed at the bottom of the
page as a way to accomplish its main missions. The document can itself be
read as a kind of ordering, as an effort to organize and clarify, or as a way of
establishing a hierarchy among values and goals, missions and concrete actions that all correspond to specific groups, with overlaps between goals; it is
a way to illustrate social action, to map an organization.

Inside the first mission, three important action values are at the core: "to
acquire, disseminate and transfer" technologies and knowledge, plus a fourth
one, "to improve" plant productivity, which implies an economic value, a reference to crop productivity. Within this vast grouping, some people do molecular biology research, but not all and not the majority, and people might
have different perceptions about this and other recent fields, for instance
biological pest control. In its generality, the document does not express these
particular fields of research with their characteristics. Researchers, who are
also seen as agents of communication and "transfer," are actors in the object domain plant productivity. There is also a place for actors "in the field"
that will be helped by the Sève network. Actors in science are connected with
potential users, producers, beneficiaries; the document communicates their
importance, but not as recipients or as publics of that research. The main
mission is seen "within a context that values the sustainability of ecosystems

and food security": We understand that here, the values of the research network have to encounter the values of a social setting. We can wonder if there
is imagined a social milieu in which, regrettably, sustainability might not be
as valued as it is by people in the Centre themselves. Or we could alternatively
say that this expresses the way by which the Centre is itself a promoter of
sustainability values and aims to further them in society, a society that does
recognize those values to a certain degree.

By being placed at the beginning, this first mission seems to be more important than the second one, which is about training qualified personnel,
but in that same paragraph we also find goals that explicitly have to do with
the exploitation and development of "non-forestry plant resources." While
excluding forestry, the phrasing expresses a relative decentering: it is the non-
forest plant life that also has to be enlarged! Do we have here anthropocen-
tric and ecocentric needs expressed as going in the same direction? It might
also be the need to enlarge available crops for different, eventually competing needs. And the second paragraph started with a first level of hierarchy,
talking about personnel to be trained, which then is met with a second one,
developing the crop resources themselves. Are the second paragraph's "ultimate" goals "more important" than the preceding elements, mentioned in
the second mission at the beginning of the same paragraph? Probably not; the
ambiguity here has more to do with ongoing organization processes and flexible hierarchy than it has to do with fixity and absolute clarity. As for the third
paragraph, we find the classical distinction between the fundamental and the
applied domains of research, mention of the molecular and the ecosystemic,
transfer issues, technological issues, and even decision-making issues. This
is a very broad description of clusters of action that actually exist inside the
Centre.

As is common in such documents, the second part of the page (after the
preceding three paragraphs) lists a series of specific, more operational aims
that seem to be secondary in comparison with the missions enunciated at
the beginning. This list gives a better description of what is actually being
done in the Centre. Among them, new elements appear that have to do with
social integration between the researchers and universities themselves. Transfer to stakeholders and to the population is mandatory from the start for
such a network to obtain funds in the first place, so it had to be mentioned
here. Outside what has already been said about training qualified personnel, the rest is about communication within research communities and for
the general public, in terms of knowledge transfer rather than in terms of
"broader impact," a set of criteria that has become very important in the
United States.

What is required in those terms is only a platform of visibility for whoever is interested, which is probably the best that can be done in such circumstances. One could surmise that people who are directly interested in that kind of very specific research might already be informed of its existence in particular instances. It certainly looks like official communication, aimed in the end at people most of the time already in the relevant network of actors who therefore can be easily joined by only pointing them in that direction. Of course, elsewhere on the website, much more information is given in terms of the members, the specific teams, current projects, ways to obtain financing from the Centre, publicity for conferences ahead or seminars, and all kinds of scientific information that cannot be looked at in detail here. Very different orientations coexist in the Centre's network, on which focus is not at all given in the mission document; for instance, investing time and energy to develop biological agriculture or instead looking into transgenesis techniques and what comes next. Said otherwise, the generality of the mission statement avoids concrete description of specific groups with their research objectives. It also does not have to discuss the relative merits, for instance of biological pest control versus research that can require specific techniques of genome or other biological material modifications.

To conclude our case study, it seems important to recall that this is science communication: a Centre communicating its core missions to a given set of recipients, to its audience, or even to their public. It is thus selecting elements that are quite general and common to everyone. It can be read as an ethical piece if we follow the fourth meaning-in-use of the word "ethics" presented before, but we do not find here debate about this or that behavior according to which norms and with what ethics officers. There is also no contrasting between possible value orientations. Nonetheless, if ethics have to do first with values at the source of human action, such a document seems relevant. It also partially shows the complexity of the network of actors involved, even though knowledge of the field might be required to better understand the whole. It says what is to be expected and even demanded, since these actors are financed by publicly funded governments and supported by their respective universities and research centers.

Discussion

In science communication as in anywhere else, ethical reflection should not be limited to problems of right-or-wrong behavior. Questions of values are also involved in human action, which has to be acknowledged as playing an important role. Some elements will have to be placed in the foreground, but actors need to be conscious of their priorities and choices. Values can be in situations of tensions, but these tensions can be lived with, distributed in a complex set in an "organized" kind of way, as was seen in our case study. One classical way to resolve potential conflicts of values is to distribute them, in specializing human action, attributing some action to some value; of course this does not work in all cases.

People have a tendency to reduce the ethical discussion to the legal or quasi-legal dimension, following or requiring the availability of norms and experts. Even though the legal dimension certainly has to be taken into account, obviously most of the time law formulation and enactment does not precede but follows developments in social life, which is foremost animated by values, involvement of people with cherished elements. For instance, we communicate science because we love science. We are interested in science because we know it is useful, it serves our purposes, it fulfills our needs, at least in some cases. It certainly has limits and can have detrimental side effects, but it is still much more efficient than other means most of the time. Science is more plural than we care to admit, and critical thinking cannot be forgotten in the purposes of science communication. The same goes for expertise issues; no amount of theoretical developments in ethics will replace the moral life of individuals and social groups, human implication in themes and discussions, even if personal intuitions have to be debatable and publicly justifiable. And it would certainly be impossible and not desirable to put everything in lawful form if we are to keep some dose of creativity and freedom in the domain of science communication. We also have to forego attempting to account for the unpredictable.

If meanings-in-use have to be taken into consideration in a very serious manner while having to deal with questions and issues, they are not sufficient even though they might come from very illustrious or famous persons and groups. We should be able to decode ethical formulations and interpret them adequately. A way to give life to those claims is to recognize their dimensions of value and reflection, e.g., to interpret them as requirements precisely and not as truth-formulations (James, 1891/1977). A reflexive approach in ethics implies that the very question of what ethics is must be asked again at the very beginning by actors while considering their partners, stakeholders, recipients, audiences, publics; a given meaning of the word cannot be just taken for granted as obvious and self-explanatory.

There is certainly a need to reflect more on science communication as a field of practice; the different types of scientific journalism, also kinds of professional communicators in their different contexts, might be more precisely characterized than what could be done here. We also have to inquire more into the actual practices of the scientist as a communicator. Required steps would involve documenting practices, and then clarifying what are the difficulties, the conflicts of values encountered, and the identifiable norms that could apply, and on that basis what remarks can be made from an ethical point of view. Discussions and better understandings of science communication ethics have to go back and constantly refer to the study of actual work that is being done in different professional contexts, which are thoroughly organizational and social. Certainly an important part of the basic ethical work to be done is to take into account those professional practices and document them more clearly, and on that basis confront the practices with the requirements indicated. Ethical reflection has to do with questioning systematically our practices, while leaving aside for a while the judgmental tendency that has been well documented and criticized (see Garfinkel, 1967). To do practical philosophy has to be something else than just encouraging us to continue to be judgmental dopes!

Regarding the case study, that concrete network has to do with plant production, whether it is agricultural in the traditional sense, biomass energy production, or other venues at all relevant levels of inquiry. I have tried to illustrate how value choices have to be combined in a complex way in practice, and how a resulting document illustrates organizational work, meaning that to write the piece was certainly trying to express how the organization understood what is common for its defining parts. The document presented together main missions, distributing the elements much more in terms of their relatively equal importance than in terms of clear-cut hierarchy and priorities. Even if, for obvious reasons, they are not placed in the forefront of the discussion, tensions can arise between teams, as always in science, situated in competition even inside a Centre that has furthered, more efficiently than many others, some original forms of collaboration. As we just saw, different aims and values have to be pursued in a conjoint manner, even though they might be very plural and hard to reconcile. At the level of generality that is illustrated by this kind of document, the best that can be said is that they have to be taken into account together, as was done by the document examined here.

References

Apel, K.-O. (1967). *Charles S. Peirce: From pragmatism to pragmaticism.* New York, NY: Humanity Books.

Buber, M. (1922). *I and thou.* New York, NY: Continuum.

Dewey, J. (1927). *The public and its problems.* New York, NY: Holt.

Dewey, J. (1939). *Theory of valuation.* Chicago, IL: University of Chicago Press.

Dewey, J. (2008a). Experience and nature. In J. A. Boydston (Ed.), *The later works of John Dewey, Vol. 1.* Carbondale: Southern Illinois University Press. (Original work published 1925)

Dewey, J. (2008b). Logic: The theory of inquiry. In J. A. Boydston (Ed.), *The later works of John Dewey: Vol. 12.* Carbondale: Southern Illinois University Press. (Original work published 1938)

Dilthey, W. (2002). The formation of the historical world in the human sciences. In R. Makkreel & F. Rodi (Eds.), *Wilhelm Dilthey: Selected works, Vol. 3.* Princeton, NJ: Princeton University Press. (Original work published 1910)

FRQNT. (2012). Centre Sève, general presentation of the missions and partners involved. [In French]. Retrieved from http://www.frqnt.gouv.qc.ca/documentsPublications/pdf/2008/RS/Centre%20SEVE-12_fevrier_08.pdf

Garfinkel, H. (1967). *Studies in ethnomethodology.* Englewood Cliffs, NJ: Prentice Hall.

Habermas, J. (1982). *On the logic of the social sciences.* Cambridge, MA: MIT Press. (Original work published 1967–1972)

Habermas, J. (1997). *Between facts and norms.* Cambridge, MA: MIT Press.

Honneth, A. (1994). *Kampf um anerkennung: Zur moralischen grammatik sozialer konlfikte.* Frankfurt, Germany: Suhrkamp.

Isaacs, W. (1999). *Dialogue and the art of thinking together.* New York, NY: Currency.

James, W. (1977). The moral philosopher and the moral life. In J. J. McDermott (Ed.), *The writings of William James* (pp. 610–629). Chicago: University of Chicago Press. (Original work published 1891)

Legault, G.-A. (1998). *Professionalisme et deliberation éthique.* Sainte-Foy, France: Presses de l'Université Laval.

Létourneau, A. (2005). Les significations majeures du mot "éthique" dans les journaux québécois, 2000–2003. *Communication, 24*(1), 177–208. Retrieved from http://communication.revues.org/3305

Létourneau, A. (2007). L'intervention en éthique: Au-delà des modèles. Chaire d'éthique appliqué, Université de Sherbrooke, Essais et conferences, 33.

Létourneau, A. (2012). Towards an inclusive notion of dialogue for ethical and moral purposes. In F. Cooren & A. Létourneau (Eds.), *(Re)presentations and dialogue* (pp. 17–36). Dialogue Studies, 16. Amsterdam, Netherlands: John Benjamins.

Létourneau, A. (2014). Perspectives d'une recherche spécifique en philosophie pratique. In A. Lacroix (Ed.), *Quand la philosophie doit s'appliquer* (pp. 151–179). Paris, France: Hermann.

Mead, G.-H. (1967). *Mind, self, & society: From the standpoint of a social behaviorist.* Chicago, IL: University of Chicago Press.

Peirce, C. S. (2001). Some consequences of four incapacities. In J. Buchler (Ed.), *Philosophical writings of Peirce.* New York, NY: Dover. (Original work published 1867)

Perelman, C., & Olbrechts-Tyteca, L. (1988). *Traité de l'argumentation: La nouvelle rhétorique.* Brussels, Belgium: Éditions de l'Université de Bruxelles.

Putnam, L. L., & Holmer, M. (1992). Framing, reframing, and issue development. In L. L. Putnam & M. E. Roloff (Eds.), *Communication and negotiation* (pp. 128–155). Newbury Park, CA: Sage.

Rawls, J. (1993). *Political liberalism.* New York, NY: Columbia University Press.

Ricoeur, P. (1981). *Hermeneutics and the human sciences: Essays on language, action and interpretation*. J. B. Thompson (Ed. and Tr). Cambridge, UK: Cambridge University Press.

Tindale, C. W. (1999). *Acts of arguing: A rhetorical model of argument*. New York, NY: SUNY Press.

Weber, M. (1992). *Essais sur la théorie de la science*. Paris, France: Plon. (Original work published 1905–1914)

14

How Discourse Illuminates the Ruptures between Scientific and Cultural Rationalities

CYNTHIA-LOU COLEMAN

In this chapter I explore how the ways we discuss scientific controversies often lead to a dualism—where *Science* gets juxtaposed against beliefs, religion, ethics, or other perspectives that are characterized as *non-Science*—in social discourse. Those of us who study how scientific issues are communicated often turn to discourse—particularly news reports, fiction books, films, television, works of art, and public displays (such as museums)—to discover how the acts of communication reveal cultural underpinnings. Such revelations are germane to the focus of this book because cultural underpinnings are composed of our values and ethics, and thus frame our perspectives and judgments.

Social discourse is considered a battlefield, according to philosopher Michel Foucault, who contends that individuals and agencies engaged in conflicts navigate their claims through all sorts of discourses—including science communication: "We are talking about a battle—the battle knowledges are waging against the power-effects of scientific discourse" (1997, p. 12). Foucault adds that, in discourse, "power and knowledge are joined together. And for this very reason, we must conceive of discourse as a series of discontinuous segments" from a variety of actors vying to establish their views as essential (1978, p. 100).

Foucault contends that discourse illuminates our worldviews. Taken from the German word *Weltanschauung*, worldviews are more than mere perspectives. Worldviews encompass what we know, how we know what we know, and how we talk about what we know—what we consider our epistemologies or ways-of-knowing—the overarching landscape that includes our values (social beliefs) and ethics (principles that govern behavior).

I am especially interested in how ways-of-knowing are framed in discourse when scientific controversies arise in matters that impact indigenous peoples. My research explores how differences in worldviews result in a dualism in discourse. For example, when covering a court ruling in North America where scientists were given permission to study indigenous bones—against the wishes of local Indian tribes—a reporter framed the ruling as an "epic struggle between science and religion" where "science won" (Westneat, 2004). When Western Science is pitted against indigenous knowledge systems in discourse, Occidental perspectives are framed as "correct," "right," and "rational," while local knowledges are often disparaged as irrational. In this vein, philosopher Michel Foucault regards indigenous epistemologies as subjugated: They are given short shrift when pitted against Occidental scientific worldviews, and, it is through discourse that meanings are created: discourse joins knowledge with power. Therefore, when knowledge systems are subjugated, they are considered "naive knowledges, located low down on the hierarchy, beneath the required level of cognition or scientificity" (Foucault & Gordon, 1980, p. 82).

Because Occidental knowledge systems ground their legitimacy in objective science, subjugated knowledges—such as indigenous worldviews—are considered rife with values whereas scientific worldviews are value-free, as evidenced in discourse. But the "value-free" perspective is a fraud, critics charge. All communication is "value laden," according to two communication theorists who write about epistemology and ethics: The Occidental approach is regarded as "cognitively clean as mathematics, built in a linear fashion from a neutral, non-contingent starting point," say Lee Wilkins and Clifford Christians (2001, p. 3). Science, however, inevitably draws from culturally derived worldviews; hence arguments that scientific objectivity shields its practitioners from bias are false. Wilkins and Christians note that communication is a creative act, and therefore discourse constructs meaning through symbols and language, which emerge from a values-laden grounding. "Culture is the womb in which symbols are born and communication is the connective tissue in culture building," they note (p. 7). And scientific worldviews, by definition, emerge from a culture that is "decisively value centered" (p. 8). Moreover, because cultural symbols and language are enfolded with values, social discourse reveals ways-of-knowing that are conjoined with our notion of ethics.

In the following pages I demonstrate how the dualism between science and non-science unfolds in discourse, presenting the ways some scientific perspectives are framed as morally correct, while some indigenous views are presented as unorthodox. This dualism sets the stage for an ethical discussion surrounding social discourse: What sort of ethics underpins discourse? I borrow my definitions from bioethicist David B. Resnik, who considers ethics "norms for conduct" and as a "method, procedure, or perspective for deciding how to act and for analyzing complex problems and issues" (2011). In the case of news discourse, reporters attend to norms that encourage "objective" reporting, which emerges from a formula that attempts to balance dueling perspectives. As Reese (1990) has observed, however, the balance imperative restricts coverage to a limited range of sources, often in a binary fashion. As a result, a multiplicity of sources is rarely gathered in the news process, and thus efforts to engage in pluralistic coverage are thwarted (Wilkins & Christians, 2001). Ethics that drive journalistic coverage therefore suffer from what Foucault and Gordon (1980) considered the cleavage of knowledge with power: Subjugated knowledges are often absent in discourse, and when they are acknowledged, they are often disparaged.

To illustrate the theories and assumptions surrounding epistemologies, I offer a case study where rationalities collide in the events and struggle surrounding the discovery of Kennewick Man in the State of Washington, and then illustrate how such rationalities unfold in discourse. I show how Occidental scientific pursuits influence the ways that indigenous people and ways-of-knowing are characterized, and present how such scientific perspectives disparage Native American knowledge systems. The news coverage surrounding the discovery of a rare 9,000-year-old skeleton in North America illuminates how the ruptures between scientific and cultural rationalities get framed in social discourse. Let me begin with another story that demonstrates how Occidental science subjugates native ways-of-knowing.

The Ice Maiden

Imagine the year is 1995. You were raised in a Peruvian village where your Incan ancestors believed they needed to assuage the mountain deities, so they sacrificed young women to Mt. Ampato. Although times have changed dramatically since Spain invaded Peru 500 years ago and conquered the Incan people, some of your indigenous relatives and friends feel their traditions have been mocked and usurped by colonizers who claimed your country for themselves. Mt. Ampato is considered sacred even by residents who have been educated in Western universities and who hold modern views buttressed by scientific reasoning. The month is September, the tail end of the Peruvian winter, when the weather is dry and cool, and you hear on the radio that an American anthropologist found a partly buried body on Mt. Ampato. The body is dug out of the melting ice and named "Juanita, the Ice Maiden"—a well-preserved Incan mummy offered up to the mountain by her village. Cur-

rent news frames the body as a "treasure" characterized as "a perfectly pre-
served" Incan sacrifice (Puffer, 2000). Juanita is pored over by experts while
crews film the threads of her clothing, the curl of her lips, and the wrinkles
of her face. Magazine stories, television documentaries, news interviews, and
books soon follow the discovery of the mummy. Juanita is then taken on the
road for public view. Encased in a glass freezer, she travels to North America.
A reporter writes that upon viewing the exhibit, President Bill Clinton joked,
"If I were a single man, I might ask that mummy out. That's a good-looking
mummy" (Kennedy, 1996).

Critics charge that the public display of ancient bodies offers a "cult of
remembrance." In their book, *Reading National Geographic*, Lutz and Collins (1993) wonder if our fascination with preserving ancient bodies sig-
nifies "imperialist nostalgia" that allows modern viewers and readers to
mourn "the passing of what we ourselves have destroyed" (p. 97). Indeed,
some anthropologists—like Johan Reinhard, who dug out Juanita from the
mountain—engage in a "race" against time to "save" the dead, study them,
and mount them for exhibit. Such displays of the dead "fetishize" the body,
according to Wakeham (2008). If we think of a fetish as an object that sym-
bolizes some desire, then Juanita's body serves as a signifier of the racialized
Other, adds Wakeham.

Not surprisingly the issue of "race" underpins much of the criticism of
grave robbing and public presentation of bodies. Discovery of the Ice Maiden
was funded by *National Geographic*, whose images of denizens of the Third
World in native dress and undress capture them in stasis, offering readers the
image of what Edward Said famously called "The Other" (1978). Such discoveries are justified as scientific pursuits. In her critique of museums, Wakeham
writes that "branches of technoscience prey upon the remainders of bodily
decay, labor to fetishistically reconstruct so-called lost precontact authenticity, and reinscribe taxonomies of otherness" (p. 166). Wakeham captures the
essence of much of the criticism of racism that is disguised as scientific pursuit. Historian James Clifford takes the critique one more step, noting that
creation of The Other separates "us" from "them" in ways that are more than
cultural or geographic: The cleavages are inspired by epistemology and methodology (Clifford, 1988, pp. 143–144).

If Clifford is correct—if scientific pursuits of the racialized Other, bol-
stered by scientific methods, enable "us" to distinguish ourselves from
"them"—then the discovery of ancient bones would necessarily leverage Occidental science over subjugated knowledge systems. Moreover, questions of
the ethics of such pursuits and their methodologies—unearthing of bodies,
for example—are warranted because they are enacted from what Resnik

considers an objective and unbiased approach: what I call "sciencing"—the act of conducting science. Foucault would likely regard sciencing as the juncture of knowledge-building with power: the act of research, and the point in time when a scientist makes a judgment that impacts The Other, as in the case when health researchers recruited African American veterans in Tuskegee, Alabama, for treatment of diseases including syphilis, and failed to follow through on treatment for many of the men who had the disease ("Tuskegee Timeline," 2016).

Theoretical Underpinnings

We now turn to the theoretical foundation that grounds the dualism of Occidental and indigenous epistemologies. Rationality is pivotal when examining how divergent worldviews are characterized in discourse. Philosopher Jürgen Habermas noted that Occidental approaches to science assume a Cartesian lens that assists scientists in imposing a mantle of objectivity that shields them from accusations of infusing values in their judgments. René Descartes sketched out a logical and rational framework to characterize the study of knowledge (epistemology) where the object of study is separated from the individual. The division between subject and object is a critical component of Western rationality, where truth is said to exist regardless of human values. In short, knowledge and truth are testable (Coleman & Herman, 2010).

Habermas places scientific reasoning and rationality within the context of *modernity*, which refers to knowledge systems beginning with the Enlightenment—a period of "modernism" arising in the wake of Isaac Newton's influential *Principia* and the French Revolution of the eighteenth century. This rationality is clearly Occidental, in the sense that its underpinnings emerge from Western European epistemologies.

Scientific rationality, therefore, resides in modernity, characterized by an objective assessment of the world buttressed by empiricism, or, knowing through evidence. But Habermas acknowledges different dimensions of rationality, including a non-rational, pre-rational, or mythical rationality, which he speaks of without derision. Moreover, modernist rationality springs from histories of non-rationality, so, in a sense, they are inseparable. Dallmayr explains Habermas' views as a sense that "vaunted rationality of modernity can be said to rest itself on non-rational presuppositions in the sense that it cannot be rationally redeemed or sustained in a comprehensive and non-circular (or non-tautological) manner" (1988, p. 571). Dallmayr further notes that scientific rationality presents a slanted view; that is, "the one-sided emphasis on cognitive-scientific and instrumental rationality to the detriment of more 'communicative' modes of reasoning; instead, the source lies in the one-sided stress on formal rationality per se" (p. 571). James Clifford takes up the gauntlet of "communicative modes of reasoning" by noting that, for collectives without written languages, "reality is rooted in oral encounter" (1988, p. 258). For American Indians, "most of what is central to their existence is never written" (p. 340). The text, therefore, becomes an important marker of the empirical and the rational. Non-written forms of communication, such as oral narratives, hold less currency with empiricists.

Such views of reasoning mesh well with studies of social discourse. From Habermas' perspective, a communicative act is one in which an actor "seeks *rationally* to *motivate* another by relying on the illocutionary binding/bonding effect (*Bindungseffekt*) of the offer contained in the speech act" (Habermas, 1990/2001, p. 58, italics in original). As communication scholars, we would therefore expect to find nuggets of meaning within social discourse, which is aimed at convincing audience members of the validity of the claims made. An important corollary to Habermas' notion of the communicative act is suggested by Willkins and Christians, who contend that communication is more than binding, bonding, and ideological: communication is persuasive (2001, p. 2). Thus, we can examine discourse by digging more deeply into knowledge systems and ask how rationalities unfold.

But first I would like to show how rationalities lead a double-life that is often contradictory. Habermas' dialectic of rational and non-rational (or, mythical) worldviews, forces some of us—journalists and scientists, for example—to split off science from non-science in our discussions. But what happens when we try to build bridges between the dualism of science and non-science? Alonzo Plough and Sheldon Krimsky tried to build bridges in their analysis of risk assessment and risk communication. Plough and Krimsky (1987) separated the technical assessment of risks from a cultural assessment, noting that each perspective is borne from rational thought. By labeling such perspectives as *scientific rationality* and *cultural rationality*, we refrain from considering one perspective as purely rational and the other as lacking rationality (Coleman, 1994). As Plough and Krimsky note, "cultural reason does not deny the role of technical reason; it simply extends it. The former branches out, while the latter branches in" (1987, pp. 8–9). Branching out refers to considering perceptions of risk within their social context to better understand public sentiment, while branching in refers to more narrow and reductionist approaches to understanding data. By viewing ways-of-knowing through both prisms—narrow and broad—we get a richer and more inclusive perspective. And, speaking from an ethical perspective, as scholars, we should attempt to bridge rationalities.

Communication Methodologies

How should we approach the investigation into social discourse of scientific and cultural rationalities? My methods employ deep readings and viewings of text and artifacts surrounding scientific controversies in Indian Country. My objective is to make sense of the meanings of the texts and messages in social discourses about scientific controversies. The approach is arguably subjective, so, to raise your confidence in my interpretations, I use what literary scholar Jonathan Culler (1997) calls a "complex investigation of meaning" through an analysis in which the researcher examines and compares multiple opportunities of mediated social discourse, such as books, news reports, and films, with the hope of uncovering expressions of rationalities. I therefore look at how rationalities are characterized as mythical, ethical, religious, cultural, scientific, and more.

To enrich my methods of discourse analysis, I consider Kristeva's (1980) concept of "intertextuality," where one medium's narrative is reflected (and thus transformed) by another medium. For example, recall the creature brought back to life by Victor Frankenstein in Mary Shelley's 1818 novel. What comes to mind is a pastiche of images: Boris Karloff as the monster in the 1931 moving picture, Gene Wilder as Dr. Frankenstein in Mel Brooks' 1974 comedy, and Aaron Eckhardt in the 2014 apocalyptic film, *I, Frankenstein.* Shelley's monster takes yet another shape in television cartoons for children (*Frankenstein Jr. and The Impossibles*), on breakfast cereal packages (*Frankenberry*), and in popular music lyrics (with songs about Frankenstein sung by musicians ranging from Sam Cooke to Weird Al Yankovic). Intertextuality also arises when discussing genetically modified crops, where critics employ the monster metaphor of Frankenfoods in popular media stories.

Such examples illustrate how one literary text *(Frankenstein)* has been twisted and transformed within myriad mediums: referring back to (self-referencing) the original creator, Frankenstein, and his monster, while simultaneously creating what Baudrillard (1994) would regard as a forgery of the original creation: The contemporary monster bears little resemblance to the original Frankenstein—who was the demon's creator.

In writing about intertextuality, Porter (1986) argues that discourse—in addition to reflecting the past—is a function of its context, and that texts are woven by threads of meaning borrowed from a variety of contexts and histories. Plough and Krimsky's work aligns with this approach as witnessed in their assessment of worldviews, noting that scientific perspectives often ignore the social context of a conflictual issue—one that might induce fear and uncertainty among publics. That such fears may not be "rational" is hardly the point: the fears are still real. Quoting Mary Douglas, Plough and Krimsky note that "acceptable standards of morality and decency" are fundamental to cultural rationality (Douglas, as cited in Plough & Krimsky, 1987, p. 8). While an engineer might argue that hydraulic fracturing (fracking) is unlikely to pollute a community's groundwater, local residents may remain skeptical—not because they lack scientific reasoning, but because they may find the engineer's reassurances immoral and indecent.

Just as the study of human perception of science and risk is best understood within a social context, the analysis of the context in which messages are conveyed is equally important. Porter offers a quote from literary scholar Vincent B. Leitch that "the text is not an autonomous or unified object, but a set of relations with other texts. Its system of language, its grammar, its lexicon, drag along numerous bits and pieces—traces—of history so that the text resembles a Cultural Salvation Army" (Leitch, as cited in Porter, 1986, p. 41). "Intertextuality," Porter writes, "animates all discourse" (p. 34).

The study of discourse, then, requires a sort of "tacking back and forth" between what myriad forms of discourse report about rationalities (Turner, 2014). Theorists argue that messages derive their meanings from a web of texts (intertextualities) and a variety of contexts (civil wars, technological inventions, and deathly plagues, to name three). As a result we need to be mindful of the ways in which meanings are constructed when analyzing messages that invoke scientific and cultural empiricism.

Science communication, Foucault and Gordon (1980) assert, pivots on the conceit of biopolitics, so any examination of modernist empiricism must add a layer of biology-cum-politics. According to Foucault, biopolitics shape the way we talk, think, and engage in questions regarding "life." Andrée put it this way:

> Biopolitics refers to the particularly modern relations of power, rooted in specific expert truth-claims and material practices that enable the regulation and efficient production of "life" by scientists, governments and industries, as well as the forms of resistance that emerge in this context. (2002, p. 168)

By ladling biopolitics onto our discussion, we must consider how decisions about life *and* death, and indigenous mummies and bones, are explained in text. We should also view the ways judgments (truth-claims and validity-claims) are constructed and then opposed.

One salient example of a biopolitical examination of discourse is the news coverage surrounding the case of Theresa Schiavo, a woman left in a "persistent vegetative state" after suffering loss of oxygen to her brain in 1990 (Coleman & Ritchie, 2011). Schiavo's husband requested life support withdrawn from Theresa in 1998. The case became fodder for pundits and politicians worldwide, and reflected a biopolitical frame in philosopher Slavoj Žižek's musings that, while "tens of millions" of individuals are "dying of AIDS and hunger all around the world," popular discourse reduced the complex issues surrounding life and death to one single example (Žižek, 2005).

I weave Foucault's sense of biopolitics with Habermas' notion of "validity claims" to help uncover how myriad rationalities from Plough and Krimsky's pursuits infuse intertextually in social discourse. When a reporter remarks after a landmark court ruling that "science won" in the "epic struggle between science and religion," he acknowledges the rupture between rationalities as a long-fought struggle in which one perspective holds sway over another (Westneat, 2004). That science "won" illustrates not only the victory over se-curing a court judgment—it also signals the progressive nature of science, where old paradigms are successfully replaced by new worldviews (see, for example, Kuhn, 1962).

For our purposes, then, we can think of discourse as animated, and lay-ered intertextually—activating meanings from a variety of sources, histories, politics, perspectives, contexts and rationalities. With this in mind, the chap-ter examines the framing of scientific issues that impact indigenous peoples, with the objective of illuminating the ways that scientific and cultural ratio-nalities are invoked in the case study of Kennewick Man.

The Kennewick Man Discovery and Discourse

On a warm summer day in 1996, two college students waded through the Co-lumbia River to sneak into the annual boat races in Kennewick, Washington. One of them stumbled over what turned out to be a human skull. Local offi-cials were notified, and law enforcement officers extracted a nearly complete skeleton from the river. Soon after, authorities contacted a consultant with a doctorate in anthropology, who was asked to examine the remains. James Chatters quickly set to work: He made an impression of the skeleton and skull, and then sent off a chunk of bone for carbon dating. Meantime, offi-cials who oversee the Columbia River and the protection of natural resources, including members of the Umatilla Tribe, were called on to discuss how to manage the human remains. The Umatilla argued that the skeleton should be returned to tribes under the Federal provision NAGPRA (the Native Amer-ican Graves Protection and Repatriation Act). The Federal law was passed in 1990 with the purpose of providing a "process for museums and Federal agencies to return certain Native American cultural items—human remains, funerary objects, sacred objects, or objects of cultural patrimony[1]—to lineal descendants, and culturally affiliated Indian tribes and Native Hawaiian organizations" (National Park Service, n.d.).

A key rationale for passage of the NAGPRA legislation was to afford American Indians a legal platform to secure artifacts that had been stolen or unearthed, including skulls and bones. Chatters worried Kennewick Man would be turned over to the tribes without giving scientists time to study the remains. He wrote, "I had been given a gift from the past, an opportunity to learn from an ancient ancestor and to convey what he could tell us for future generations" (2002, p. 78). Chatters contacted fellow anthropologists, including those at the Smithsonian Museum of Natural History, and a group was formed that sued for the right to study the "gift."

Carbon dating revealed the skeleton was between 8,000 and 10,000 years old: a rare discovery. Chatters met with reporters to talk about the skeleton, now dubbed "Kennewick Man" by news reporters and called "The Ancient One" by tribal elders. According to a story in *The Washington Post,*

> Chatters declared that the skeleton "looks like no one I've ever seen before." But if he had to choose a category, he would say the bones looked "Caucasoid," most resembling those of a "pre-modern European." (Coll, 2001)

The descriptor "Caucasoid" was seized by members of the press corps, who quickly replaced the scientific terminology with "Caucasian." A *New York Times* reporter wrote: "From head to toe, the bones were largely intact. The skeleton was that of a man, middle-aged at death, with Caucasian features, judging by skull measurements" (Egan, 1996).

By framing Kennewick Man as Caucasian, news reporters fashioned a new lens to view the skeleton, and asked: If the skull is Caucasian, does that mean Kennewick Man was "White"? While anthropologists argued over human migration theories and the morphology of skulls, Chatters again looked to news media to announce he found a perfect doppelgänger for Kennewick Man: an English actor named Patrick Stewart, whom Chatters claimed resembled the ancient denizen. Television programs and magazines placed photographs of the contemporary actor famous for his portrayal of a science-fiction starship captain from the future (Jean-Luc Picard) side-by-side with a clay bust of Kennewick Man.

---

1. Note that cultural patrimony refers to an object that has "ongoing historical, traditional, or cultural importance central to the Native American group or culture itself, rather than property owned by an individual Native American," according to the NAGPRA Glossary (National Park Service, n.d.).

Some called Chatters' claims merely speculative, and one scholar asked why it was necessary to label the skull Caucasoid and make "controversial and incendiary efforts" (Coll, 2001). An Australian broadsheet waxed whimsical, musing, "We can only presume that, in a plot worthy of *Star Trek: The Next Generation*, Captain Picard somehow managed to travel back in time to North America, where he fathered the entire local Indian population" (McDonald & Yeaman, 1998).

Speculation the skeleton was "White" increased in public discourse after Chatters' comments. As a result, American Indians seeking the return of the bones were accused of political trickery because they wanted the skeleton reburied, as custom and tradition dictate. But Chatters told *60 Minutes* reporter Leslie Stahl that local tribes were afraid—afraid that Kennewick Man would upset Native narratives. "They have a history now, the way it's laid out, that fits their present-day political needs quite effectively. If that history changes, it may not fit so well" ("Profile," 2002).

Stahl concluded that the tribes were fearful that scientists would discover Kennewick Man was not a Native American, and, "if someone else was here before they were, their status as sovereign nations and all that comes with it—treaty rights and lucrative casinos, like this one on the Umatilla Reservation—could be at risk." The charge that American Indians would lose their sovereignty and resources—if Kennewick Man's bones were deemed different from modern-day tribal people—weighs heavy with irony for indigenous residents of the Pacific Northwest. The use of "sovereignty" in Indian Country (arising from the French term for "ruler") springs from the notion that indigenous tribes signed agreements with officials representing the United States government that Native Americans would relinquish their lands and natural resources in exchange for the right to self-govern (sovereignty). Thus, Kennewick Man's origin—if ever determined—would likely have no effect on sovereignty, making the political attacks specious.

From a Foucauldian lens, the linkage of politics with self-determination reveals the power struggles between the scientists who wanted to study the skeleton and the Indians who wanted The Ancient One returned. The scientific investigation became elevated over indigenous concerns when biopolitical views were infused in the discourse. One scientist feared that lawyers would "use the results out of context" in order to win a judgment for the Native tribes: "Their goal is to win an argument," the scientist told the newspaper. "My goal is to find the truth by scientific experiments" ("Expert: Radiocarbon dating complex, uncertain," 1999). While Native American advocates were described as politically motivated, scientific pursuits were often framed as free of politics, agendas, and values. Referring to conducting tests on the skeleton, one advocate noted, "they are doing science, so the scientific position is winning time after time" ("Umatillas blast plans for DNA tests," 2000).

News coverage of the skeletal discovery has flowed and ebbed since 1996, swelling from time to time as news about court judgments made headlines. Initially the courts ruled in favor of the Federal government in order to protect Indian remains and return them to local tribes. In turn, the scientists who sued to study the bones argued that tribes could not demonstrate the skeleton was culturally affiliated with local tribes, as required by NAGPRA. Tribes responded that proving affiliation after the passing of thousands of years was unreasonable, and that the only conclusion, based on a cultural rationality, was that the skeleton was indigenous to the territory. The legal case rested on who gets to define what "Native American" and "cultural affiliation" mean. Pacific Northwest Indians link their ways-of-knowing with oral stories and with the experiences of living in the place where their epistemologies and ontologies arise. The origins of The Ancient One are, by definition, native to place, thus addressing the definition of "Native American" and "cultural affiliation." But in the hands of the non-Indian, the definitions of key terms are stripped of their cultural underpinnings. Language varies according to a society's interests, mused anthropologist Claude Levi-Strauss. Naming something is a function of the power to label, and thus reflects cultural values. In short, language constructs our realities (Mileaf, n.d.).

Gordon M. Sayre, who studies Native American representations in literature, argues that, in the case of Kennewick Man, Indian tribes were required to act in accordance with para-legal and quasi-scientific definitions replete with jargon, and, as a result, tribes were unable to muster evidence "to meet the statutory requirement of a 'cultural affiliation' linking the bones to any local Indian tribes" (2005, p. 52). Sayre asks, "Can any culture, anywhere in the world, prove that its language, its material culture, and its name for itself, have not changed in 9,000 years?"

To dig more deeply into the news coverage and language resulting from the discovery of The Ancient One, my colleague Erin Dysart Hanes and I examined framing in news discourse from 1996 to 2004 (Coleman & Dysart, 2005). We found much of the news coverage linked the scientific judgments with political ones, thus embracing what Foucault would consider biopolitical framing. For example, Leslie Stahl's *60 Minutes* report landed on what she considered the heart of the conflict: "I'm asking about the central question of concern: Who was here first?" ("Profile," 2002).

Stahl is not the only journalist to cement reportorial framing around provenance. *The New York Times* and *Newsweek* ask, "Who got here first?" (Wilford, 1999; Murr, 1999). The *London Independent* wonders, "Who are the true Native Americans?" (Carlin, 1996), and *Canada's Globe and Mail* claims, "Skulls enliven debate on earliest Americans" (McIlroy, 2003). In other words, much of the discourse settled within the domain of the scientists engaged in the conflict over handling the remains, who successfully framed the issue as a landmark discovery that only "sciencing" could solve.

Indians were portrayed in discourse as superstitious, regressive, and antiscience, as Dysart and I noted in our analysis of news coverage. Returning the bones to Indians was described as a "head-in-the-sand attitude of a preliterate society" and "bad science" (Coleman & Dysart, 2005, p. 19).

Steve Coll, managing editor for *The Washington Post* during the landmark case, writes, "From the start, the scientists suing over Kennewick Man have shaped public perception of the case by emphasizing their free speech arguments and painting the Indians as irrational and superstitious" (Coll, 2001).

Local tribes, another reporter charged, wanted The Ancient One returned in order to conceal evidence of the body's origins. "The Indians want, quite literally, to bury it," noted John Carlin of the *London Independent* (Carlin, 1996). "New evidence suggests that the Indians are not quite as indigenous as previously believed, that they themselves might be descended from the same stock as their erstwhile European tormentors," Carlin wrote.

Coverage most frequently became embroiled in war-like terms, often as a battle between scientists and Indians, science and religion, scientific inquiry and tribal rights, reason and superstition, and cowboys and Indians (Coleman & Dysart, 2005, p. 14). Stories took on the trope of warpaths, conflict, and conflagration, thus echoing Foucault's contention that discourse is a battlefield.

Kennewick Man Is, After All, Native American

Legal wrangling over the ancient remains continued until 2004, when the courts determined that the tribes failed to demonstrate a cultural linkage with the skeleton. Scientists were given the nod to examine the bones, and public discourse slowed to a trickle, until an exciting discovery was announced a decade later, in May 2014.

Researchers claimed they had linked ancient DNA to Native Americans, a decade after the Kennewick Man court case ended. Like a scene taken from an Indiana Jones movie, scientists found a young female skeleton trapped in an underwater cave in the Yucatan peninsula. The team named the 12,000-year-old skeleton, "Naia," after the Greek water nymphs, and extracted a small sample of her DNA. Her genetic material revealed a surprise: Naia's DNA matched that of modern-day American Indians, according to a report published in *Science* in May 2014. The findings would prove pivotal for arguments that contemporary indigenous Americans are genetically linked to ancient Americans—including Kennewick Man (Balter, 2014).

Moreover, the researchers said Naia's cranium resembles that of Kennewick Man's: "A long and high skull and pronounced forehead. These differ markedly from those of today's Native Americans, who tend to have broader and rounder skulls," according to Michael Balter of *Science*. Naia's profile chalks up additional evidence in favor of the hypothesis that modern-day Native Americans descended from early Paleoamericans, who some Western scientists believe migrated to the Americas across the Bering Strait. Naia's discovery signals an important paradigm shift for theories underpinning the origins of Kennewick Man.

In fact, the lead author on the study is James Chatters, who famously referred to Kennewick Man's skull as Caucasoid years earlier and told the press The Ancient One bore more resemblance to an English actor than to contemporary American Indians. But Chatters corrected his earlier mistakes, taking a "personal about-face," Balter noted. "Now, he [Chatters] fully embraces the genetic continuity model" (2014).

While new evidence of America's peopling flowed through anthropology circles, it barely caused a ripple in popular news, despite the evidence that Kennewick Man's and Naia's skeletal features shot holes in the presumption that, because the ancient skulls *looked* different than those of modern-day Indians, they were presumed to be genetically different. Not so, according to the 2014 *Science* article.

It would take another 13 months for news reporters to make a connection between Kennewick Man and present-day Native Americans. It turns out scientists in Denmark had been working prodigiously to decipher the genetic material extracted from Kennewick Man, and they announced in June 2015 that new technology finally allowed them to analyze the skeleton's DNA. Scientists said they had linked Kennewick Man's genetic profile with members of the Colville tribe in Washington and that they had finally found the link between modern denizens and The Ancient One (Rasmussen et al., 2015).

News headlines announced the DNA discovery as evidence that Kennewick Man is genetically affiliated with contemporary Indians: "DNA shows 8,500-year-old Kennewick Man was Native American" (*USA Today*); "DNA confirms Kennewick Man's genetic ties to Native Americans" (National Public Radio); "DNA reignites Kennewick Man debate" (BBC News); "DNA proves Kennewick Man, the Ancient One, is native; Tribes continue fight for reburial" (*Indian Country Today*); "DNA analysis proves Kennewick Man is Native American" (CBS News); "Case closed? DNA links 8,500-year-old

Kennewick Man to Native Americans" (NBC News); "New DNA results show Kennewick Man was Native American" (*The New York Times*); and "First DNA tests say Kennewick Man was Native American" (*The Seattle Times*).

The stories focused on the breakthroughs that enabled scientists to unravel the ancient DNA, but only a few news articles delved deeply into the conflict that arose from competing rationalities resulting from distinctly different cultural worldviews. Reporters instead embraced what Tuchman (1973) called the "what-a-story" frame, where the genetic discovery solved the mystery over whether Kennewick Man was indigenous. More thoughtful coverage about the ethical scope of dueling epistemologies was largely absent.

One notable exception is coverage by Native American news networks. Indian Country Today Media Network interviewed Jim Boyd, chairman of the Colville Confederated Tribes, who said the scientists' early claims that Kennewick Man was unrelated to modern Indians were "wrong." And it makes you wonder, Boyd said, "What else is wrong?" The erroneous conclusion that Kennewick Man was European, he added, had been published and "entered people's belief systems . . . and that's dangerous" (Taylor, 2015).

Armand Minthorn, the Umatilla elder who was embattled in the lawsuits with anthropologists who sued to study Kennewick Man, spoke at a press conference following the announcement of the Danish study's results. Minthorn reminded reporters of his public stance from two decades earlier: "We have always said the same thing since 1996. This individual is Native American. Period. End of discussion. These remains need to be treated in a sacred manner. These remains are sacred to us" (Burke Museum video, 2015).

Why Rationalities Matter

I have argued that social discourse—especially news coverage—focuses our attention on the characterization of scientific conflicts. Examining the Kennewick Man story through a news lens magnifies the drama that pits dueling ideologies on what Foucault called the battlefield of discourse. The bulk of coverage unfolds in a biopolitical vein, where the constructs of "life" and "death" are debated. The very meanings of what constitute life and death—and how they should be "managed"—reveal fundamentally different worldviews emerging from scientific and cultural rationalities. When we burrow into the layers of meaning we can see ethical threads running throughout discourse. Resnik (2011) reminds us that scientific ethics stem from norms of behavior and objectivity embedded within and expected from scientific practice, or "sciencing." In other words, these ethics cling to the notion of "doing good science."

But "doing good science" is not enough when American Indian communities question the motivation for such practices. In his collection of science perspectives from American Indian writers, Keith James (2001) laments the idea that science is value-free: "Scientists and engineers often invoke this ideal as a talisman to confer a veil of sanctity on their work despite abundant evidence that the human mind, even a scientist's, is inherently subjective in all its operations" (p. 48).

When the Havasupai Nation of the Southwestern United States sued scientists for failing to fully disclose their reasons for studying tribal members' blood—as required by ethical standards when studying human subjects—one geneticist defended herself in *The New York Times*, saying, "I was doing good science" (Harmon, 2010). Native American scholars Vine Deloria Jr. and Daniel Wildcat (2001) extend "sciencing" into a discussion of morality and ethics, asserting that "Western science has no moral basis" (p. 4). Deloria and Wildcat write that ethics are infused within American Indian epistemologies, in part because "all relationships are ethical" and because relationships among humans, animal creatures, the earth, the skies, streams, and oceans undergird ways-of-knowing (p. 27). Deloria and Wildcat note that, because scientific rationalities claim to be value-free, they are also devoid of meaning. Meaning, they write, is derived from our ethical relationships with the world (pp. 32–33).

In the case of Kennewick Man, the Federal legislation designed to protect Indian remains was neutered by a legal system that, in managing life and death, and their attendant meanings, required Native communities to provide evidence that Kennewick Man was an ancestor, assigning the burden of proof to the tribes and demonstrating a biopolitical interpretation of NAGPRA. Scientists had no such requirement to offer up evidence, and the search for meaning (*What does it mean to be Native American?*) was reduced to the practice of sciencing, affording the anthropologists and the scientific rationality worldview a heightened legitimacy. As Foucault has suggested, the undercurrent of biopolitics essentially frames the arguments over who controls and manages the meanings of life and death. For the Native Americans who demanded the return of the 9,000-year-old ancestor, their contention is that the ancient bones are sacred, and thus require great care. "Sacred human remains are not artifacts. They are what they are—sacred—and they are our ancestral remains, and they need to be treated as such," Armand Minthorn told journalists ("Bones of contention," 2001). Kennewick Man is regarded as a relative, not a pile of old bones.

For many Native Americans, the ethical entailments of conducting scientific procedures on their relatives (whether they are living or dead) gets nearly lost in the discourse over who may claim provenance over skeletal remains and for what purposes. Sciencing, by its fundamental nature, embodies values surrounding the progressive trajectory of discovery: Juanita, Naia, and Kennewick Man are heralded as buried treasures that deserve study, in the style demanded by scientific rationality. On the other hand, claims made by American Indians are swept aside like dust that has settled on more than 18,000 Native American remains housed at the Smithsonian Institution. The specimens were collected from "graveyards and military battlefields in the American West during the 19th century, as the Army waged what amounted in many cases to campaigns of extermination against indigenous tribes," according to *The Washington Post* (Coll, 2001).

One of the anthropologists who sued to study the bones of Kennewick Man simply dismissed the Indian claims, saying, "The fact is that, hey, there was a big war. The world has had a lot of these issues of conquered peoples, and you know, one doesn't like that sort of thing, but that's the reality. It happened." The speaker, Robson Bonnichsen, effectively frames the Indians as regressive: "Can we resurrect and make history right? I don't think so. . . . I mean, hey, life goes on."

In summary, the case study of Kennewick Man shows how social discourse reveals ways that meanings are constructed that emerge from different ways-of-knowing and how their ethical underpinnings are fundamentally different. In the instance of scientific rationality, ethics are linked to the practice of science, whereas Native American cultural rationalities conjoin ethics within ontological relationships. In other words, "sciencing" becomes an ethical act regardless of the consequences ("I was doing good science"). Enactments of scientific procedures—dredging Kennewick Man from the Columbia River drink and warming mummy Juanita's frozen skin in a makeshift laboratory—are, by Occidental science's definition, ethical. That some American Indian epistemologies consider relationships with the universe as the basis for ethical behavior offers a quite different perspective on ethics than does sciencing. Deloria and Wildcat draw a poignant conclusion (p. 24): "The universe is alive, and it is personal."

References

Andrée, P. (2002). The biopolitics of genetically modified organisms in Canada. *Journal of Canadian Studies, 37*(3), 162–191.

Balter, M. (2014). Bones from a watery "black hole" confirm first American origins. *Science, 344*(6158), 680–681.

Baudrillard, J. (1994). *Simulacra and simulation.* Ann Arbor: University of Michigan Press.

Bones of Contention. (2001, June 19). PBS Newshour (transcript). Public Broadcasting Service, USA. Retrieved from http://www.pbs.org/newshour/bb/science-jan-june01-kennewick_6-19/

Boyle, A. (2015, June 18). Case closed? DNA links 8,500-year-old Kennewick Man to Native Americans. NBC News. Retrieved from http://www.nbcnews.com /science/science-news/dna-links-kennewick-man-skeleton-native-american-tribes-n377291

Burke Museum. (2015, June 18). Kennewick Man (The Ancient One) press conference at the Burke Museum (video). Retrieved from https://www.youtube.com/watch?v=Px-UQ3X8UvU

Carlin, J. (1996, October 5). Who are the true Native Americans? Are red men "true" Native Americans? *The Independent* (United Kingdom).

Casey, M. (2015, June 18). DNA analysis proves Kennewick Man is Native American. CBS News. Retrieved from http://www.cbsnews.com /news/dna-analysis-proves-kennewick-man-is-native-american/

Chatters, J. C. (2002). *Ancient encounters: Kennewick Man and the first Americans.* New York, NY: Simon & Schuster.

Clifford, J. (1988). *The predicament of culture: Twentieth century ethnography, literature and art.* Cambridge, MA: Harvard University Press.

Coleman, C.-L. (1994). Science, technology and risk coverage of a community conflict. *Media, Culture and Society, 17,* 65-70.

Coleman, C-L., & Dysart, E. V. (2005). Framing of Kennewick Man against the backdrop of a scientific and cultural controversy. *Science Communication, 27*(1), 1-24.

Coleman, C-L., & Herman, D. (2010). Ways of knowing. *Smithsonian National Museum of the American Indian* (Winter), 29-33.

Coleman, C.-L., & Ritchie, L. D. (2011). Examining metaphors in biopolitical discourse. *Lodz Papers in Pragmatics, 7*(1), 29-59.

Coll, S. (2001, June 3). The body in question: The discovery of the remains of a 9,000-year-old man on the Columbia River has set off a conflict over race, history and identity that isn't just about the American past, but about the future as well. *The Washington Post,* p. W8.

Culler, J. (1997). *Literary theory: A very short introduction.* Oxford, UK: Oxford University Press.

Dallmayr, F. (1988). Habermas and rationality. *Political Theory, 16,* 553-579.

Deloria, V., Jr., & Wildcat, D. R. (2001). *Power and place: Indian education in America.* Golden, CO: Fulcrum Resources.

Doughton, S. (2015, June 17). First DNA tests say Kennewick Man was Native American. *The Seattle Times.* Retrieved from http://old.seattletimes.com /html/localnews/2025488002_kennewickdnaxml.html

Douglas, M. (1985). *Risk acceptability according to the social sciences.* New York, NY: Russell Sage Foundation.

Egan, T. (1996, September 30). Tribe stops study of bones that challenge history. *The New York Times.* Retrieved from http://www.nytimes.com/1996/09/30/us/tribe-stops-study-of-bones-that-challenge-history.html?pagewanted=all&src=pm

Expert: Radiocarbon dating complex, uncertain. (1999, December 28). *The Associated Press.*

Foucault, M. (1978). *The history of sexuality* (1st American ed.). R. Hurley (Tr.). New York, NY: Pantheon Books.

Foucault, M. (1997). Society must be defended: Lectures at the Collège de France, 1975-1976. D. Macey (Tr.), M. Bertani & A. Fontana (Eds.). New York, NY: Picador.

Foucault, M. (2004). *Power as knowledge* (3rd ed.). Boulder, CO: Westview Press.

Foucault, M., & C. Gordon (1980). *Power/knowledge: Selected interviews and other writings, 1972–1977* (1st American ed.). New York, NY: Pantheon Books.

Habermas, J. (2001). *Moral consciousness and communicative action.* C. L. Hardt & S. W. Nicholsen (Trs.). Cambridge, MA: MIT Press. (Original work published 1990)

Harmon, A. (2010, April 21). Indian tribe wins fight to limit research of its DNA. *The New York Times.* Retrieved from http://www.nytimes.com/2010/04/22/us/22dna.html?pagewanted=1&ref=health&_r=0

James, K. (2001). Culture: The spirit beneath the surface. In K. James (Ed.), *Science and Native American communities: Legacies of pain, visions of promise* (pp. 45–50). Lincoln: University of Nebraska Press.

Joyce, C. (2015, June 18). DNA confirms Kennewick Man's genetic ties to Native Americans. Health-shots radio broadcast, National Public Radio. Retrieved from http://www.npr.org/sections/health-shots/2015/06/18/415205524/dna-confirms-kennewick-mans-genetic-ties-to-native-americans

Kennedy, H. (1996, May 24). Mum's not the word for Bill. *The New York Daily News.* Retrieved from www.nydailynews.com

Kristeva, J. (1980). *Desire in language: A semiotic approach to literature and art.* New York, NY: Columbia University Press.

Kuhn, T. S. (1962). *The structure of scientific revolutions.* Chicago, IL: University of Chicago Press.

Leitch, V. B. (1983). *Deconstructive criticism.* Ithaca, NY: Cornell University Press.

Lutz, C., & Collins, J. L. (1993). *Reading National Geographic.* Chicago, IL: University of Chicago Press.

McDonald, P., & Yeaman, S. (1998, June 22). Star Trek the past generation? *The Advertiser/Sunday Mail* (Adelaide, South Australia).

McIlroy, A. (2003, September 4). Skulls enliven debate on earliest Americans. *The Globe and Mail* (Toronto, Ontario, Canada). Retrieved from http://www.theglobeandmail.com/technology/science/skulls-enliven-debate-on-earliest-americans/article18429893/

Mileaf, J. (n.d.). Janine Mileaf on Levi-Strauss, science of the concrete. Department of English, University of Pennsylvania website. Retrieved from http://www.english.upenn.edu/~jenglish/Courses/mileaf.html

Morelle, R. (2015, June 18). DNA reignites Kennewick Man debate. Science and Environment Section, BBC News. Retrieved from http://www.bbc.com/news/science-environment-33170655

Murr, A. (1999, November 15). Who got here first? The war over the first Americans rages as science sifts through spear points-and shibboleths. *Newsweek.*

National Park Service. (n.d.). National NAGPRA. Retrieved from https://www.nps.gov/nagpra/INDEX.HTM

Plough, A., and Krimsky, S. (1987). The emergence of risk communication studies: Social and political context. *Science, Technology & Human Values,*12(3&4), 4–10.

Porter, J. E. (1986). Intertextuality and the discourse community. *Rhetoric Review,* 5(1), 34–47.

Profile. (2002, September 15). *60 Minutes,* CBS News Network. CBS news transcript. Retrieved from Lexis-Nexis. Originally aired October 28, 1998, S. Finkelstein (producer), L. Stahl (reporter).

Puffer, B. (2000, November). Preserving a mummy. *NOVA.* PBS online. Retrieved from http://www.pbs.org/wgbh/nova/peru/mummies/preservehome.html

Rasmussen, M., Sikora, M., Albrechtsen, A., Korneliussen, T. S., Moreno-Mayar, J. V., Poznik, G. D., . . . & Willerslev, E. (2015). The ancestry and affiliations of Kennewick man. *Nature, 523,* 455–458.

Reese, S. (1990). The news paradigm and the ideology of objectivity: A socialist at the *Wall Street Journal. Critical Studies in Mass Communication, 7*(4), 390–409.

Resnik, D. B. (2011; updated 2015). What is ethics in research and why is it important? National Institute of Environmental Health Sciences. Retrieved from http://www.niehs.nih.gov/research/resources/bioethics/whatis/index.cfm

Said, E. W. (1978). *Orientalism.* New York, NY: Pantheon.

Sayre, G. M. (2005). Prehistoric diasporas: Colonial theories of the origins of Native American peoples. In P. D. Beidler and G. Taylor (Eds.), *Writing race across the Atlantic world, medieval to modern* (pp. 51–75). London, UK: Palgrave Macmillan.

Shelley, M. W., & Butler, M. (1994). *Frankenstein, or, the modern Prometheus: The 1818 text.* Oxford, UK: Oxford University Press.

Taylor, K. (2015, June 18). DNA proves Kennewick Man, the Ancient One, is native: Tribes continue fight for reburial. *Indian Country Today Media Network.* Retrieved from http://indiancountrytodaymedianetwork.com/2015/06/18/dna-proves-kennewick-man-ancient-one-native-tribes-continue-fight-reburial-160780

Tuchman, G. (1973). Making news by doing work: Routinizing the unexpected. *American Journal of Sociology, 79*(1), 110–131.

Turner, J. (2014). *Philology: The forgotten origins of the modern humanities.* Princeton, NJ: Princeton University Press.

Tuskegee Timeline. (2016, January 7). U.S. Public Health Service syphilis study at Tuskegee. Centers for Disease Control and Prevention website. Retrieved from http://www.cdc.gov/tuskegee/timeline.htm

Umatillas blast plans for DNA tests. (2000, February 2). Associated Press. *The Columbian* (Vancouver, WA).

Wakeham, P. (2008). *Taxidermic signs: Reconstructing aboriginality.* Minneapolis, MN: University of Minnesota Press.

Watson, T. (2015, June 18). DNA shows 8,500-year-old Kennewick Man was Native American. *USA Today.* Retrieved from http://www.usatoday.com/story/news/2015/06/18/kennewick-man-skeleton-dna/28927311/

Westneat, D. (2004, July 21). We all own Kennewick Man's bones. *The Seattle Times.* Retrieved from http://community.seattletimes.nwsource.com/archive/?date=20040721&slug=danny21

Wilkins, L., & Christians, C. (2001). Philosophy meets the social sciences: The nature of humanity in the public arena. *Journal of Mass Media Ethics, 16*(2/3), 99–121.

Wilford, J. N. (1999, November 9). New answers to an old question: Who got here first? *The New York Times.* Retrieved from http://www.nytimes.com/learning/general/featured_articles/991109tuesday.html

Zimmer, C. (2015, June 18). New DNA results show Kennewick Man was Native American, Science section, *The New York Times.* Retrieved from from http://www.nytimes.com/2015/06/19/science/new-dna-results-show-kennewick-man-was-native-american.html?_r=0

Žižek, S. (2005, December). Biopolitics: Between Abu Ghraib and Terri Schiavo. *Artforum.* Retrieved from https://artforum.com

Afterword

SUSANNA PRIEST, JEAN GOODWIN,
AND MICHAEL F. DAHLSTROM

We hope those who have worked through this book seeking an explicit, comprehensive, and coherent set of ethical principles applicable to the way science is communicated—whether by scientists, professional communicators, or ordinary citizens—will by now realize why this is not, for the most part, possible. Science and technology are vast and complex areas of practice that emerge at the cutting edge of human knowledge. As they emerge, some elements become controversial, while others gain or lose priority as attention shifts. Different developments may challenge human values (as well as professional practice) in different ways. New ethical issues often arise alongside new scientific developments (see chapter 11 by Arduser, this volume). And as is true for all social norms, perceptions about what responses are ethical can vary—and can themselves shift as well.

Further, one ethical ideal may interfere with the fulfillment of another. Scientists may feel more ethical obligation to keep their labs working well (and their staff and students employed) than to spend more time on public communication (see chapter 9 by Davies). They might want to support good public decision making but also have commitments to specific policy options that they feel serve the public good more than others. Journalists following their own professional norms by consulting expert sources may, even if inadvertently, create a distorted impression (see chapter 7 by Kruvand). Business people may be guided by one set of ethics, while ordinary citizens may march to a different drum—or more accurately many different drums—and yet all may be behaving in ethically defensible ways. Choosing among alternative actions often becomes a matter of perceptions and values, not simply one of rules.

The solution is not to give in to the apparent confusion, nor is it necessarily to develop a new set of rules. Indeed, communicating evidence-based risk information suggests one set of ethical considerations (see chapter 2 by Thompson), while communicating about the goals of a scientific organization might suggest another (see chapter 13 by Létourneau). It certainly isn't to conclude that ethics is irrelevant, either. Ethical principles are almost always in flux, though, and in practice may depend on many considerations of context and purpose. The solution is twofold: first, to recognize that a modern, pluralistic society supports many sorts of ethical and value systems, some of which may be in conflict at any given moment, and second, to recognize that science itself, as a set of social institutions, can embody and reflect some of these same conflicts. The only real solution is for all of us to be more thoughtful and reflective in our everyday behavior.[1]

Science communication has been going on as long as science has been going on, but as a recognized academic and professional field, it is relatively young. So far as we know, this is the first book to focus centrally on the ethics of this field, which is itself still emerging. However, the ethics of scientific publishing, the ethics of communication, including the ethics of communication professions such as journalism and of public relations (yes, public relations does have ethics!), and the ethics of scientific research itself have much deeper histories and provide rich and well-established resources on which to draw. Yet, in the end, individual science communicators are often faced with the practical necessity of making up their own minds about what is ethical in a given situation, especially as their field of practice continues to evolve—and as science itself raises new questions.

While the studies of ethics in this book draw from many disciplinary roots (from philosophy to social science) and from many forms of science (from genetics and biotechnology to climate and environmental science), each illustrates considerations that are both unique to its context and yet more general. We hope that readers will not be too tempted to avoid the discus-

---

1. Philosophers and other ethicists have long debated whether some ethical rules are indeed universal. We cannot settle this longstanding debate here. However, to those of us who are social scientists, it goes almost without saying that all such rules are social constructions—that is, they are the products of human thought and perception, at least in part. There are some near-universals; to take an obvious example, today's human societies (unlike some past societies that routinely practiced human sacrifice) generally prohibit the forceful killing of human beings—and yet in many places we still have the death penalty, and risking human lives in war can be seen as necessary for some sorts of greater good. Some, of course, dispute the ethics of both of these apparent exceptions to a "universal" prohibition against killing, but that has not made them go away.

sions not directly grounded in their own field but rather ask themselves what lessons from each chapter are relevant to their particular work. While most ethical challenges in science communication and in science itself arise in specific contexts, they can still apply to other contexts as well. And yet we do understand that some chapters will be of more interest to particular readers than others, and we have built the book to also accommodate those who do not plan to read it from beginning to end.

How Does Ethics Matter?

The central message of this book is that ethics permeates much of human decision making, even science-related decision making and even in contexts that we don't think of as involving reflection on ethics. Acting in an ethical way is actually a means of protecting our social group; we often do so, as previously pointed out in this book, without much conscious thought. Perhaps there is even a genetic foundation for this, since making ethical choices so clearly contributes to the success of human social groups, from families to entire civilizations—although so far as we know this has not been demonstrated scientifically.

As science communicators, scientists and other scholars, social scientists and ordinary citizens, we usually behave ethically without thinking very much about it. That seems to get us by in most situations. As members of a very remarkable species, we routinely adapt to complicated social norms that seem constantly in flux. However, scientists' commitment to the uncovering of ultimate truth, like journalists' commitment to complete objectivity, can blind people to the role of ethics and values in their own professional work, even imbuing them with a certain sense of privileged status. Conditions of rapid change or active conflict sometimes cause the grounding of our choices in ethics and values to suddenly become more visible and emerge as topics of discussion and even active contention (see, for example, chapter 14 by Coleman).

This book has also highlighted a few of the areas where scientists and science communicators, even those consciously adhering to established principles for good behavior, can act unethically without intending or realizing it. If we could always "trust our gut"—that is, follow our intuitive sense of right and wrong—life would in some ways be easier. It is both healthy and helpful to pay attention to those intuitions. Who among us has not observed or engaged in some act and only realized first through some kind of visceral internal emotional response (anger or guilt or shame) that that act was simply "wrong"? Importantly, this takes place without any necessity to consult a rule list. Yet the existence of different moral compasses that guide different individuals in different situations, especially in a shrinking globe inhabited by populations of increasing cultural diversity, may challenge our intuitive reactions.

Social scientific studies of how science communication influences its audiences—a big part of what is today sometimes called the "science of science communication"—are powerful. But this book has sought to show how other disciplines and other considerations matter as well, both to further our understanding of how ethics pervades our professional work and to inform effective science communication practice. Science writers often make a choice between whether to present the science they write about as being divorced from particular scientists as people or personalize it instead by respectful communication about the scientists who produced that science (see chapter 6 by Ranalli). Such choices can affect audience interpretations of character and motive; so can a reporter's inclusion of information designed to help readers sort out fact from interpretation or opinion (see chapter by 10 McKaughan and Elliott).

Science as Power

This discussion would be incomplete without bringing in one final consideration: All players are not equal in terms of power. This makes science political, in part. It could hardly be otherwise. Thus major international seed companies may have more power than individual local farmers (see chapter 12 by Bronson), and major international energy companies may have more influence than local environmental groups (see chapter 8 by Gaither and Sinclair). Typically, scientists have more power (in the form of credibility) than advocacy groups who may question some of their conclusions. Some forms of delivery (such as through the use of narrative—see chapter 5 by Dahlstrom and Ho—or featuring powerful sources or, increasingly, utilizing so-called social media) may have more power, or at least a different kind of power, than traditional journalism. Some communication "frames" (key ideas or associations suggested by messages) seek to empower people to make decisions (see chapter 4 by Sprain), while others may subtly disempower them by promoting particular strategic interests (see chapter 3 by Priest). And then, of course, there is the power of actual politicians who may want to fund some science (perhaps including scientific work being done in their own districts) but not other science (including science with which they or their constituencies disagree, or at least devalue).

Yet some scientists can and do have opportunities to take control over how their purposes and motives are understood. Myriad new models are being advanced that put scientists into direct contact with citizens, such as science cafés, citizen science projects, town hall discussions, science festivals, and many other formats. Interactive museum exhibits may not feature interaction with actual scientists but are designed to encourage cognitive engagement even so. Entertainment television shows that make science fun may not provide direct contact between scientists and audiences, but they offer opportunities for vicarious contact, enabling audience members to identify with scientists and see them as actual people.

Research is needed to more firmly establish whether these various formats actually shape attitudes or simply entertain audiences, but it seems evident that at least some learning and attitude formation do go on through them. Such events can have the effect (often deliberate) of changing the roles of the scientists who participate in them from researcher, lecturer, or other impersonal and powerful "ivory tower" role to a more nearly coequal one, whether as conversational partner or informational resource or work-team member. This role shift tends to change a lot of things, from the point of view of both communication and social psychology (including the inferred intent of whatever involved scientists might say or do; see chapter 1 by Goodwin). Yet, even in a dialogic setting, the participants in a conversation are not necessarily equal. Experiments in public engagement with science can be motivated by ethical considerations, chiefly concern over how to better practice democracy in the creation of science policy, but are unlikely to erase the underlying power differentials between experts and others.

It is our hope through this book to make scientists and science communicators more thoughtful about how ethics matters to their work. Most scientists are committed to science as a career because they are committed to the value of science itself, and most science communicators have much the same commitment. Social norms, however, evolve. We have moved beyond an era in which science is generally accepted without question—and this is not necessarily a bad thing from the point of view of the health of democracy. Yet there are still material differences between science and pseudo-science and between mainstream and "maverick" science. Our evolving and ever-more-complex society needs both science and science communication; it also needs ongoing reflection on ethical considerations pertinent to each of these.

A Final Note

Of course, any book based on academic work cannot end without suggestions for future research. As this is a fledgling area, it is difficult to be specific, on the one hand; yet, on the other, there are almost too many possible directions to consider, making it especially hard to think about which ones should have priority. Every chapter in this volume could easily have its own proposed list of follow-up questions to be answered.

One particular challenge of this interdisciplinary area is that representatives of multiple disciplines really do all bring unique perspectives that represent vital contributions to the collective whole. How could we engage in the analysis of science communication ethics without people who understand the science itself, without people who understand science communication studies, or without people experienced in ethical analysis? As Rachelle Hollander, former National Science Foundation program officer and subsequent director of the Center for Engineering Ethics and Society at the National Academy of Engineering, put it in a conversation with one of us many years ago, this cannot be accomplished solely by more people becoming interdisciplinary in their approaches to their own individual work. Rather, people with deep knowledge of the relevant disciplinary academic work also need to be willing to put in the extra effort to communicate and collaborate across disciplines about their discipline-specific work.

All three perspectives—and in some cases more—are necessary. Just as many chapters in this volume sprang from work discussed at interdisciplinary workshops at Iowa State University, the context and opportunity for that cross-disciplinary communication must often be specially created. While some fortunate individuals may find enough like-minded scholar-researchers within their own institution, for many this interaction will take place only as the result of an extensive networking and grant-seeking effort. It will often involve travel, which of course costs money. We believe the results are well worth that effort, but academic institutions may not always reward this work in proportion to its logistical difficulty and its actual real-world importance.

Not surprisingly, given the interdisciplinary nature of this field, quite a few of the possible future research directions we can imagine are also interdisciplinary: What ethical challenges face today's scientist-activists hoping to influence both public opinion and public policy by their actions? Are there more areas than "greenwashing" or the sale of home genetics tests where the ethics of advertising and marketing could conflict with the ethics of other science communication contexts, and how do we resolve those conflicts? What kind of individual public presence should scientists and other science communicators seek to create and maintain—for example, through blogs or other public statements? What ethical principles should apply to the visual representation of science and technology—whether done through new communication technologies or old ones?

This is hardly anything close to a comprehensive list—merely illustrations of how broad this territory actually is. We hope some of our readers will seek to become leaders of a new cohort of scholars for whom science communication ethics is central rather than peripheral to their work on a multitude of problems.

Contributors

LORA ARDUSER (PhD, Technical Communication and Rhetoric, Texas Tech University) is an assistant professor in the Professional Writing program at the University of Cincinnati. She has published in journals including the *Journal of Technical Writing and Communication*, *Women's Studies in Communication*, *Computers and Composition*, and *Narrative Inquiry*. Her research interests include the rhetoric of health and medicine and the health humanities. Her forthcoming book is entitled *Living Chronic: Agency and Expertise in the Rhetoric of Diabetes* (The Ohio State University Press).

KELLY BRONSON (PhD, Science and Technology Studies, St. Thomas University) is a Canada Research Chair candidate in Science and Society at the University of Ottawa in Canada. She is a social scientist researching governance of controversial technologies such as GMOs, big data, and hydraulic fracturing. Her research aims to bring community values and environmental demands into conversation with technical knowledge in the production of evidence-based decision-making. She has published her work in regional (*Journal of New Brunswick Studies*), national (*Canadian Journal of Communication*), and international (*Journal of Responsible Innovation* and *Big Data & Society*) journals.

CYNTHIA-LOU COLEMAN (PhD, Journalism and Mass Communication, University of Wisconsin–Madison) is a professor of communication at Portland State University. She studies the social construction of science in mainstream discourse and the effects of news framing on policies that impact American Indian communities. She has held fellowships with the Centers for Disease Control and Prevention and the Smithsonian National Museum of the American Indian, and her work has been published in journals such as *Journal of Health and Mass Communication*, *Journalism and Mass Communication Quarterly*, *American Studies/Indigenous Studies Today*, and *Science Communication*.

MICHAEL F. DAHLSTROM (PhD, Journalism and Mass Communications/ Environmental Resources, University of Wisconsin–Madison) is an associate professor and the associate director of the Greenlee School of Journalism and Communication at Iowa State University. Dahlstrom's research focuses on the effects of narratives on perceptions of science, and his research has been published in journals such as *Science Communication, Communication Research,* and *Media Psychology.* Dahlstrom served as the head of the Communicating Science, Health, Environment and Risk Division of the Association for Education in Journalism and Mass Communication in 2015–2016.

SARAH R. DAVIES (PhD, Social Studies of Science, Imperial College London) is assistant professor and Marie Curie Research Fellow in the Department of Media, Cognition and Communication, University of Copenhagen. Her research interests include public understanding of science, deliberative democracy, and studies of emerging technologies. Recent publications include articles in *Social Studies of Science* and *Science and Technology Studies,* and the books *All Hackers Now* (Polity, 2016) and *Science Communication: Culture, Identity and Citizenship* (Palgrave, 2016).

KEVIN C. ELLIOTT (PhD, History and Philosophy of Science, Notre Dame) is an associate professor in Lyman Briggs College, the Department of Fisheries and Wildlife, and the Department of Philosophy at Michigan State University. His research lies at the intersection of the philosophy of science and practical ethics. He is the author of *Is a Little Pollution Good for You? Incorporating Societal Values in Environmental Research* (Oxford, 2011) and *A Tapestry of Values: An Introduction to Values in Science* (Oxford, 2017), as well as numerous journal articles and book chapters.

BARBARA MILLER GAITHER (PhD, Journalism and Mass Communication, University of North Carolina at Chapel Hill) is an associate professor in the School of Communications at Elon University. Her research concerns the messaging that companies and industry trade groups use to influence public opinion, policy debates, and/ or legislative outcomes. She is concerned with how audiences react to and are influenced by corporate advocacy messages and corporate social responsibility campaigns. Her work has recently been published in *Social Marketing Quarterly, Environmental Communication,* and the *Journal of Applied Communication Research.*

JEAN GOODWIN (PhD, Rhetoric, University of Wisconsin–Madison) is a professor in the Communication Department at North Carolina State University, where she is affiliated with the Leadership in Public Science cluster. She also holds a JD degree from the University of Chicago Law School. Her work is in rhetoric, focusing on civic argumentation and in particular on the communication of science in policy controversies. Recent publications have appeared in journals such as *Argumentation* and *WIREs Climate Change.*

SHIRLEY S. HO (PhD, Communication, University of Wisconsin–Madison) is an associate professor and assistant chair (faculty) in the Wee Kim Wee School of Communication and Information at Nanyang Technological University, Singapore. Her research focuses on understanding the underlying mechanisms behind public perception of science, health, and environmental issues. Her work has appeared in premier journals such as *Science Communication, Public Understanding of Science,* and *Nature Nanotechnology.* She is an associate editor for the *Asian Journal of Communication, Environmental Communication,* and the *Oxford Encyclopedia for Climate Change Communication.*

MARJORIE KRUVAND (PhD, Journalism, University of Missouri) is associate professor in the School of Communication at Loyola University Chicago. Previously, she was a journalist, a public relations professional, and a Knight Science Journalism Fellow at MIT. Her research focuses on health and science communication, mass communication of bioethical issues, the use of expert sources, and dynamics between public relations professionals and journalists. Her work has been published in such journals as *Case Studies in Strategic Communication, Science Communication, Public Relations Review,* and the *Journal of Mass Media Ethics.*

ALAIN LÉTOURNEAU (PhD, Philosophy and History of Religions, Université Paris-IV Sorbonne and Institut Catholique de Paris) is a professor in the Department of Philosophy and Applied Ethics at the Université de Sherbrooke (Quebec, Canada). He has worked for 15 years on questions of ethics, communication, and environmental governance. He has published numerous journal articles and is author, editor, or coeditor of eleven books, notably *Validité et Limites du Consensus en éthique* (L'harmattan); he also contributed to a 2014 volume on *Philosophy of Communication Ethics,* edited by Ronald Arnett and Patricia Arneson.

DANIEL J. MCKAUGHAN (PhD, History and Philosophy of Science, Notre Dame) is associate professor in the Philosophy Department at Boston College. He is coeditor, with Holly Vande Wall, of *History and Philosophy of Science: A Reader* (Bloomsbury, 2017). McKaughan's interests include philosophy of science, philosophy of biology, epistemology, and philosophy of religion, and his work has been published in a diverse range of journals including *Isis, Philosophy of Science, Biology and Philosophy, Studies in History and Philosophy of Science, Journal of Philosophical Research, Religious Studies,* and *Science.*

SUSANNA PRIEST (PhD, Communications, University of Washington) has held tenured faculty appointments at Texas A&M University, the University of South Carolina, and the University of Nevada, Las Vegas over the past 25 years. She is also a Fellow of the American Association for the Advancement of Science. Her research focuses on public responses to emerging technologies. She is now an independent scholar, author, and consultant, and she edits the journal *Science Communication*. She has published in journals ranging from *Critical Studies in Mass Communication* to *Risk Analysis*, and her most recent book is *Communicating Climate Change: The Path Forward* (Palgrave Macmillan, 2016).

BRENT RANALLI (MSc, Environmental Science and Policy, Central European University) is an independent scholar with the Ronin Institute and a policy professional at The Cadmus Group. His scholarly work focuses on scientific norms, virtue ethics, and cultures of environmental sustainability. The recipient of a research fellowship from the Thoreau Society, he coedits *Environment: An Interdisciplinary Anthology* and his work has appeared in *Spontaneous Generations*, the *Kennedy Institute for Ethics Journal*, and the *Journal of Sustainability Education*, among other publications.

JANAS SINCLAIR (PhD, Mass Communication, University of Florida) is an adjunct associate professor in the School of Media and Journalism at the University of North Carolina at Chapel Hill. Her research focuses on the psychology of persuasion for messages about science, technology, and the environment and the basis of lay public attitudes on these topics. Her research has been published in major advertising, mass communication, and science communication journals including the *Journal of Advertising*, *Journalism and Mass Communication Quarterly*, and *Risk Analysis*.

LEAH SPRAIN (PhD, Communications, University of Washington) is an assistant professor in the Department of Communication at the University of Colorado Boulder. Her research focuses on democratic engagement, including deliberation, engagement across difference, and environmental governance. Outreach and praxis are crucial to democratic engagement; thus, much of her work is focused on the practice-theory interface. Her work has appeared in *Communication Theory*, *Journal of Applied Communication Research*, and *Environmental Communication*, and she coedited *Social Movement to Address Climate Change: Local Steps for Global Action*.

PAUL B. THOMPSON (PhD, Philosophy, State University of New York at Stony Brook) is a professor of philosophy and the W. K. Kellogg Chair in Agricultural, Food and Community Ethics at Michigan State University. His research has centered on ethical and philosophical questions associated with agriculture, especially the environmental and food safety risks associated with the production and consumption of food. His most recent book, *From Field to Fork: Food Ethics for Everyone*, was published by Oxford University Press in 2015.

Index

Made in the USA
Lexington, KY
15 November 2018